ENERGY

SHORTAGE, GLUT, OR ENOUGH?

ISSN 1534-1585

ENERGY
SHORTAGE, GLUT, OR ENOUGH?

Sandra Alters

INFORMATION PLUS® REFERENCE SERIES
Formerly published by Information Plus, Wylie, Texas

GALE®

THOMSON
—————★—————™
GALE

Detroit • New York • San Diego • San Francisco • Cleveland • New Haven, Conn. • Waterville, Maine • London • Munich

Energy: Shortage, Glut, or Enough?

Sandra Alters

Project Editors
Kathleen J. Edgar and Ellice Engdahl

Editorial
Paula Cutcher-Jackson, Debra Kirby, Prindle
LaBarge, Charles B. Montney, Heather Price

Permissions
Debra Freitas

Product Design
Cynthia Baldwin

Composition and Electronic Prepress
Evi Seoud

Manufacturing
Keith Helmling

For permission to use material from this product, submit your request via Web at
http://www.gale-edit.com/permissions, or you
may download our Permissions Request form
and submit your request by fax or mail to:

Permissions Department
The Gale Group, Inc.
27500 Drake Rd.
Farmington Hills, MI 48331-3535
Permissions Hotline: 248-699-8006 or
800-877-4253, ext. 8006
Fax: 248-699-8074 or 800-762-4058

Cover photograph reproduced by permission of
PhotoDisc.

Since this page cannot legibly accommodate all
copyright notices, the acknowledgments consti-
tute an extension of the copyright notice.

LIBRARY OF CONGRESS CATALOGING-IN-PUBLICATION DATA

ISBN 0-7876-5103-6 (set)
ISBN 0-7876-6070-1
ISSN 1534-1585

Printed in the United States of America
10 9 8 7 6 5 4 3 2 1

TABLE OF CONTENTS

Energy is essential to human existence. The need for ever-increasing amounts of energy has resulted in some of the most profound social changes in human history. This chapter explores some of those changes, as well as current and future trends in energy production and consumption, public policy, and environmental concerns.

This chapter focuses on one of the planet's most important natural resources: oil. The discussion touches on many different aspects of the oil issue, including domestic and international production and consumption, price and demand trends, and environmental concerns stemming from the 1989 *Exxon Valdez* oil spill off the Alaskan coast.

Because of its relatively low cost and high efficiency, natural gas is becoming an increasingly important source of energy. This chapter covers the processes involved in producing, storing, and delivering natural gas; current trends in production and consumption; and predictions about the future of the gas industry.

Coal first rose to prominence as an energy source in the 19th century. After a period in which it was supplanted by oil as the world's primary energy source, coal regained some of its popularity during the oil crises of the 1970s and early 1980s. Besides tracing coal's rise and fall, and rise again, the chapter also discusses the different types and classifications of coal; mining methods; trends in price, production, and consumption; and the numerous environmental concerns that surround the use of this popular fossil fuel.

Few technological advances have inspired as much hope, and dread, as the introduction of nuclear power. As discussed here, nuclear energy offers significant advantages over fossil fuels, but there are also huge—and potentially deadly—downsides, such as the tricky propo-

sition of disposing of radioactive waste; the enormous environmental impact; and the risk of nuclear-plant meltdowns, such as the Chernobyl disaster of 1986. The future of nuclear power, including the prospects for nuclear fusion, is also discussed here.

Development of a renewable energy source—one that can be used over and over again, is economical, and does not generate any form of pollution—has long been a dream of scientists, and, as discussed here, that dream is getting closer to becoming a reality. Alternative energy sources, such as hydropower, solar and wind energy, geothermal power, and hydrogen, are all attractive options to fossil-fuel sources, but they are not without their disadvantages. This chapter details some of these alternative sources; trends in their use; and the advantages and disadvantages of each.

Because oil, gas, coal, and uranium are nonrenewable resources, it is important to know how much of each is recoverable, or potentially recoverable, from the earth. This chapter addresses that issue, focusing on domestic and international reserves of these resources; worldwide trends in the exploration for them; and the environmental impact of that exploration.

This chapter focuses on another major energy source: electricity. Among the topics discussed are domestic and international production and consumption of electricity; the deregulation of electric utilities in the United States; and projected trends in the domestic electric industry.

Public health is inextricably linked to the health of the environment. And the health of the environment is largely determined by how we use, and dispose of, our energy sources. This chapter analyzes this complex web, focusing on global warming, manufacturers' attempts to build more efficient automobiles and appliances, and projected trends in energy conservation.

PREFACE

Energy: Shortage, Glut, or Enough? is the latest volume in the ever-growing *Information Plus Reference Series*. Previously published by the Information Plus company of Wylie, Texas, the *Information Plus Reference Series* (and its companion set, the *Information Plus Compact Series*) became a Gale Group product when Gale and Information Plus merged in early 2000. Those of you familiar with the series as published by Information Plus will notice a few changes. Gale has adopted a new layout and style that we hope you will find easy to use. Other improvements include greatly expanded indexes in each book, and more descriptive tables of contents.

While some changes have been made to the design, the purpose of the *Information Plus Reference Series* remains the same. Each volume of the series presents the latest facts on a topic of pressing concern in modern American life. These topics include today's most controversial and most studied social issues: abortion, capital punishment, care for the elderly, crime, health care, energy, the environment, immigration, minorities, social welfare, women, youth, and many more. Although written especially for the high school and undergraduate student, this series is an excellent resource for anyone in need of factual information on current affairs.

By presenting the facts, it is Gale's intention to provide its readers with everything they need to reach an informed opinion on current issues. To that end, there is a particular emphasis in this series on the presentation of scientific studies, surveys, and statistics. These data are generally presented in the form of tables, charts, and other graphics placed within the text of each book. Every graphic is directly referred to and carefully explained in the text. The source of each graphic is presented within the graphic itself. The data used in these graphics are drawn from the most reputable and reliable sources, in particular from the various branches of the U.S. government and from major independent polling organizations.

Every effort was made to secure the most recent information available. The reader should bear in mind that many major studies take years to conduct, and that additional years often pass before the data from these studies are made available to the public. Therefore, in many cases the most recent information available in 2003 dated from 2000 or 2001. Older statistics are sometimes presented as well, if they are of particular interest and no more-recent information exists.

Although statistics are a major focus of the *Information Plus Reference Series* they are by no means its only content. Each book also presents the widely held positions and important ideas that shape how the book's subject is discussed in the United States. These positions are explained in detail and, where possible, in the words of those who support them. Some of the other material to be found in these books includes: historical background; descriptions of major events related to the subject; relevant laws and court cases; and examples of how these issues play out in American life. Some books also feature primary documents, or have pro and con debate sections giving the words and opinions of prominent Americans on both sides of a controversial topic. All material is presented in an even-handed and unbiased manner; the reader will never be encouraged to accept one view of an issue over another.

HOW TO USE THIS BOOK

The United States is the world's largest consumer of energy in all its forms. Gasoline and other fossil fuels power its cars, trucks, trains, and aircraft. Electricity generated by burning oil, coal, and natural gas—or from nuclear or hydroelectric plants—runs America's lights, telephones, televisions, computers, and appliances. Without a steady, affordable, and massive amount of energy, modern America could not exist. This book presents the latest information on U.S. energy consumption and production, compared with years past. Controversial issues such as the U.S. dependence on foreign oil; the possibility of exhausting fossil fuel supplies; and the harm done to the environment by mining, drilling, and pollution are explored.

Energy: Shortage, Glut, or Enough? consists of nine chapters and three appendixes. Each of the major elements of the U.S. energy system—such as coal, nuclear power, renewable energy sources, and electricity generation—has a chapter devoted to it. For a summary of the information covered in each chapter, please see the synopses provided in the Table of Contents at the front of the book. Chapters generally begin with an overview of the basic facts and background information on the chapter's topic, then proceed to examine sub-topics of particular interest. For example, Chapter 7: Energy Reserves—Oil, Gas, Coal, and Uranium begins with a description of the different ways in which natural resource reserves are measured and their reliability. The chapter then moves into an examination of U.S. reserves of oil, followed by gas, coal, and uranium. Statistics on known and estimated reserves are presented, as are projections of how long these reserves will last if they continue to be consumed at current rates. Then the chapter presents similar statistics on worldwide reserves of oil, coal, gas, and uranium. Readers can find their way through a chapter by looking for the section and sub-section headings, which are clearly set off from the text. Or, they can refer to the book's extensive index, if they already know what they are looking for.

Statistical Information

The tables and figures featured throughout *Energy: Shortage, Glut, or Enough?* will be of particular use to the reader in learning about this topic. These tables and figures represent an extensive collection of the most recent and valuable statistics on energy production and consumption; for example: the amount of coal mined in the United States in a year, the rate at which energy consumption is increasing in the United States, and the percentage of U.S. energy that comes from renewable sources. Gale believes that making this information available to the reader is the most important way in which we fulfill the goal of this book: To help readers understand the topic of energy and reach their own conclusions about controversial issues related to energy use and conservation in the United States.

Each table or figure has a unique identifier appearing above it, for ease of identification and reference. Titles for the tables and figures explain their purpose. At the end of each table or figure, the original source of the data is provided.

In order to help readers understand these often complicated statistics, all tables and figures are explained in the text. References in the text direct the reader to the relevant statistics. Furthermore, the contents of all tables and figures are fully indexed. Please see the opening section of the index at the back of this volume for a description of how to find tables and figures within it.

In addition to the main body text and images, *Energy: Shortage, Glut, or Enough?* has three appendices. The first is the Important Names and Addresses directory. Here the reader will find contact information for a number of organizations that study energy. The second appendix is the Resources section, which is provided to assist the reader in conducting his or her own research. In this section, the author and editors of *Energy: Shortage, Glut, or Enough?* describe some of the sources that were most useful during the compilation of this book. The final appendix is this book's index. It has been greatly expanded from previous editions, and should make it even easier to find specific topics in this book.

ADVISORY BOARD CONTRIBUTIONS

The staff of Information Plus would like to extend their heartfelt appreciation to the Information Plus Advisory Board. This dedicated group of media professionals provides feedback on the series on an ongoing basis. Their comments allow the editorial staff who work on the project to make the series better and more user-friendly. Our top priorities are to produce the highest-quality and most useful books possible, and the Advisory Board's contributions to this process are invaluable.

The members of the Information Plus Advisory Board are:

- Kathleen R. Bonn, Librarian, Newbury Park High School, Newbury Park, California
- Madelyn Garner, Librarian, San Jacinto College—North Campus, Houston, Texas
- Anne Oxenrider, Media Specialist, Dundee High School, Dundee, Michigan
- Charles R. Rodgers, Director of Libraries, Pasco-Hernando Community College, Dade City, Florida
- James N. Zitzelsberger, Library Media Department Chairman, Oshkosh West High School, Oshkosh, Wisconsin

COMMENTS AND SUGGESTIONS

The editors of the *Information Plus Reference Series* welcome your feedback on *Energy: Shortage, Glut, or Enough?* Please direct all correspondence to:

Editors
Information Plus Reference Series
27500 Drake Rd.
Farmington Hills, MI, 48331-3535

ACKNOWLEDGMENTS

The editors wish to thank the copyright holders of material included in this volume and the permissions managers of many book and magazine publishing companies for assisting us in securing reproduction rights. We are also grateful to the staffs of the Detroit Public Library, the Library of Congress, the University of Detroit Mercy Library, Wayne State University Purdy/Kresge Library Complex, and the University of Michigan Libraries for making their resources available to us.

Following is a list of the copyright holders who have granted us permission to reproduce material in Energy: Shortage, Glut, or Enough? *Every effort has been made to trace copyright, but if omissions have been made, please let us know.*

For more detailed source citations, please see the sources listed under each individual table and figure.

Alternative Fuels Data Center, U.S. Department of Energy: Table 9.8.

Centers for Disease Control and Prevention: Table 9.1.

Council on Environmental Quality: Figure 5.2.

Energy Information Administration: Figure 1.1, Figure 1.2, Figure 1.3, Figure 1.4, Figure 1.5, Figure 1.6, Figure 1.7, Figure 1.8, Figure 1.9, Figure 1.10, Figure 1.11, Figure 1.12, Figure 1.14, Figure 1.15, Table 1.1, Table 1.2, Table 1.3, Table 1.4, Table 1.5, Table 1.6, Table 1.7, Table 1.8, Table 1.9, Table 1.10, Table 1.11, Figure 2.1, Figure 2.2, Figure 2.3, Figure 2.4, Figure 2.5, Figure 2.6, Figure 2.7, Figure 2.8, Figure 2.9, Figure 2.10, Figure 2.11, Figure 2.12, Table 2.1, Table 2.2, Table 2.3, Table 2.4, Table 2.5, Figure 3.1, Figure 3.2, Figure 3.3, Figure 3.4, Figure 3.5, Figure 3.6, Figure 3.7, Figure 3.9, Figure 3.10, Figure 3.11, Figure 3.12, Figure 3.13, Figure 3.14, Figure 3.15, Figure 3.16, Figure 3.17, Table 3.1, Table 3.2, Table 3.3, Figure 4.1, Figure 4.2, Figure 4.3, Figure 4.4, Figure 4.5, Figure 4.6, Figure 4.7, Figure 4.8, Figure 4.9, Figure 4.10, Table 4.1, Table 4.2, Table 4.3, Table 4.4, Figure 5.1, Figure 5.5, Figure 5.6, Table 5.1, Table 5.2, Table 5.3, Table 5.5, Figure 6.1, Figure 6.3, Figure 6.6, Figure 6.7, Figure 6.8, Figure 6.9, Figure 6.10, Table 6.1, Figure 7.1, Figure 7.2, Figure 7.3, Figure 7.4, Figure 7.5, Table 7.1, Table 7.2, Table 7.3, Table 7.4, Table 7.5, Table 7.6, Table 7.7, Table 7.8, Table 7.9, Figure 8.2, Figure 8.3, Figure 8.4, Figure 8.5, Figure 8.6, Figure 8.7, Figure 8.8, Table 8.1, Table 8.2, Table 8.3, Table 8.4, Figure 9.1, Figure 9.2, Figure 9.6, Figure 9.7, Figure 9.8, Figure 9.9, Figure 9.10, Table 9.4, Table 9.6, Table 9.7, Table 9.9, Table 9.10.

U.S. Department of Energy: Figure 5.9, Figure 6.2, Figure 6.5.

U.S. Department of the Interior: Figure 3.8.

U.S. Department of Transportation: Figure 1.13.

U.S. Environmental Protection Agency: Figure 9.3, Figure 9.4, Table 9.2.

U.S. General Accounting Office: Figure 5.7, Table 5.4, Figure 6.4, Figure 9.5, Table 9.5.

U.S. Nuclear Regulatory Commission: Figure 5.3, Figure 5.4, Figure 5.8.

U.S. Office of Technology Assessment: Figure 8.1, Table 9.3.

CHAPTER 1

AN ENERGY OVERVIEW

Energy is essential to life. Living creatures draw on energy flowing through the environment and convert it to forms they can use. The most fundamental energy flow for living creatures is the energy of sunlight, and the most important conversion is the act of primary production, in which plants and phytoplankton convert sunlight into biomass by photosynthesis. Earth's web of life, including human beings, rests on this foundation.

— *Energy in the United States: 1635–2000,* U.S. Energy Information Administration, 2001

A HISTORICAL PERSPECTIVE

Pre–Twentieth Century

Over the centuries, people have found ways to harness energy, such as using animals to do work or inventing machines to tap the power of wind or water. The industrialization of the modern world was accompanied by the widespread and growing use of such fossil fuels as coal, oil, and natural gas.

Significant use and management of energy resulted in one of the most profound social changes in history within a few generations. In the early 1800s, most Americans lived in rural areas and worked in agriculture. The country ran mainly on wood fuel. One hundred years later, most Americans were city dwellers and worked in industry. America had become the world's largest producer and consumer of fossil fuels, had roughly tripled its per capita use of energy, and had become a global superpower.

The United States has always been a resource-abundant nation. However, it was not until the Industrial Revolution in the mid-1800s that the total work output of engines surpassed that of work animals. As the United States industrialized, coal began to replace wood as a primary fuel. Then, as industrialization proceeded, petroleum and natural gas replaced coal for many applications. The United States has relied heavily on these three fossil fuels—coal, petroleum, and natural gas.

FIGURE 1.1

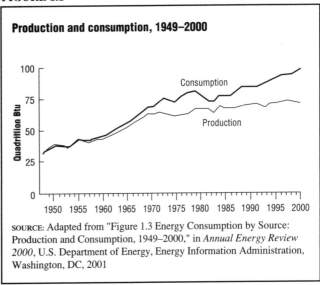

Production and consumption, 1949–2000

SOURCE: Adapted from "Figure 1.3 Energy Consumption by Source: Production and Consumption, 1949–2000," in *Annual Energy Review 2000*, U.S. Department of Energy, Energy Information Administration, Washington, DC, 2001

The Twentieth Century

For much of its history the United States has been nearly energy self-sufficient, although small amounts of coal were imported from Britain in colonial times. Through the 1950s domestic energy production and consumption were nearly equal. During the 1960s consumption slightly outpaced production. By the 1970s the gap had widened and continues to do so. (See Figure 1.1.) Since the 1970s energy imports have been used to try to close the energy production/consumption gap. However, America's dependence on other countries for some of its energy needs has brought problems.

CRISIS IN THE 1970s. In 1973 the United States supported Israel in its war with its Arab neighbors. In response, several of these Arab nations cut off exports of oil to the United States and decreased exports to the rest of the world. The U.S. embargo was lifted six months later, but the price of oil tripled from the 1973 average to

FIGURE 1.2

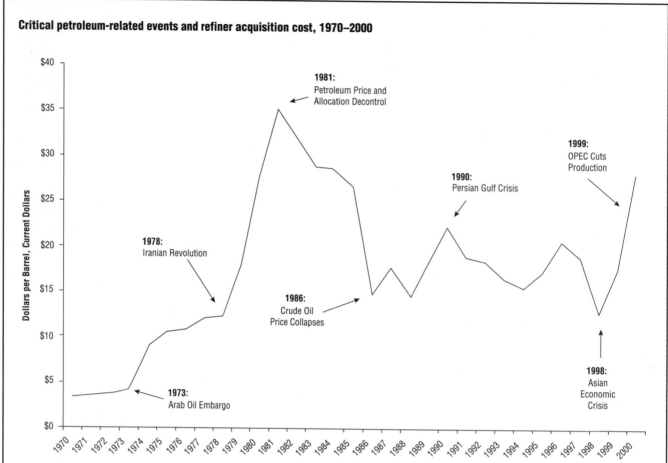

Critical petroleum-related events and refiner acquisition cost, 1970–2000

SOURCE: "Petroleum Chronology Chart: Critical Petroleum-Related Events and U.S. Refiner Acquisition Cost, 1970–2000," in *Petroleum Chronology of Events 1970–2000,* U.S. Department of Energy, Energy Information Administration, Washington, DC, 2002 [Online] http://www.eia.doe.gov/pub/oil_gas/petroleum/analysis_publications/chronology/petroleumchronology2000.htm [Accessed: October 27, 2002]

about $12 per barrel. (See Figure 1.2.) Not only did Americans (and others around the world) face sudden price hikes for products produced from oil, such as gasoline and home heating oil, but they faced temporary shortages as well. The energy problem quickly became an energy crisis, which led to occasional blackouts in cities and industries, temporary shutdowns of factories and schools, and frequent lines at gasoline service stations. The sudden increase in energy prices in the early 1970s is widely considered to be a major cause of the economic recession of 1974 and 1975.

Oil prices increased even further in the late 1970s. The Iranian revolution began in late 1978 and resulted in a significant drop in Iranian oil production from 1978 to 1981. During this same period the Iran-Iraq war began, and many other Persian Gulf countries decreased their output as well. Companies and governments began to stockpile oil. As a result, prices continued to rise. Prices were high until 1983. (See Figure 1.2.)

PRICES FALL IN THE 1980s. In early 1981 the U.S. government responded to the oil crisis by removing price and

allocation controls on the oil industry. That is, the government no longer controlled domestic crude oil prices or restricted exports of petroleum products, preferring to allow the marketplace and competition to determine the price of crude oil. Therefore, domestic oil prices rose to meet foreign oil prices.

As a result of these increasingly high prices, individuals and industry stepped up their conservation efforts and switched to alternative fuels. The demand for crude oil declined. (See Figure 1.2.) However, the Organization of Petroleum Exporting Countries (OPEC), and particularly Saudi Arabia, cut its output in the first half of the 1980s to keep the price from declining dramatically. (Current members of OPEC are Algeria, Indonesia, Iran, Iraq, Kuwait, Libya, Nigeria, Qatar, Saudi Arabia, United Arab Emirates, and Venezuela.)

In 1985 Saudi Arabia decided to increase its market share of crude oil exports by increasing its production. (Saudi Arabia is now the world's largest producer and exporter of oil and the key member of OPEC.) Other OPEC members followed suit, which resulted in a glut of crude oil on the world market. Crude oil prices fell sharply in early 1986, and imports increased.

UPS AND DOWNS IN THE 1990s AND EARLY 2000s. In August 1990 Iraq invaded Kuwait, and the public feared an oil shortfall caused by the United Nations (UN) embargo on all crude oil and oil products from both countries. Prices rose suddenly and sharply, but non-OPEC countries in Central America, western Europe, and the Far East, along with the United States, stepped up their production to fill the gap in world supplies. When the UN approved the use of force against Iraq during the Persian Gulf crisis in October 1990, prices fell quickly. (See Figure 1.2.)

The collapse of Asian economies in the mid-1990s led to a further drop in the demand for energy, and petroleum prices dipped sharply in the late 1990s. OPEC reacted by curtailing production, which boosted prices in 2000. (See Figure 1.2.) World crude oil prices then declined through 2001 as global demand dropped because of weakening economies (especially in the United States) and a drop in jet fuel demand following the September 11, 2001, terrorist attacks.

As of late 2002 the Israeli-Palestinian terrorist attacks and counterattacks resulted in concerns that Iraq might halt its crude oil shipments to countries that supported the Jewish state of Israel over Islamic Palestine. Additionally, concerns existed that the Middle East region might become destabilized should the United States invade Iraq, which has the second-largest oil reserve in the world. These were two primary factors that caused crude oil prices to rise in 2002.

GOVERNMENTAL ENERGY POLICIES

Under President Ronald Reagan

In 1977 President Jimmy Carter, a Democrat, described the energy problem of the time as one that could only be "effectively addressed by a Government that accepts responsibility for dealing with it comprehensively and by a public that understands its seriousness and is ready to make necessary sacrifices." However, when Republican Ronald Reagan took over the presidency, he downplayed the importance of government responsibility for dealing with the energy problem. The Reagan administration sharply cut federal programs for energy and opposed government intervention in energy markets. For example, the administration refused to tax energy imports, which may have stimulated domestic production and conservation. President Reagan believed that the expansion of the federal government's role in energy policy was counterproductive and misguided. His administration transferred the center of the decision-making process to the states, the private sector, and individuals.

Under President George H. W. Bush

The subsequent Republican administration of President George H. W. Bush continued the Reagan policy of limiting government regulation of the energy industry. In 1991 President Bush unveiled a long-awaited energy policy that promised to reduce U.S. dependence on foreign oil by increasing domestic oil production and the use of nuclear power. President Bush's aim was to rely on "the power of the marketplace, the common sense of the American people, and the responsible leadership of government and industry." He planned to achieve this by, among other proposals, producing additional oil from environmentally sensitive areas, encouraging pipeline construction, simplifying the construction permit process for nuclear power plants, and increasing competition in the production of electricity. His proposals did not include government-directed conservation efforts or tax incentives.

Conservationists disagreed with President Bush, objecting to increased offshore drilling, especially in the coastal plain of the Arctic National Wildlife Refuge (ANWR) in Alaska. They also wanted to see automobile fuel economy improved and conservation methods stressed, rather than increasing the use of nuclear power.

Under President Bill Clinton

The Democratic administration of President Bill Clinton brought government involvement back into energy and environmental issues, although the major energy bills proposed by this administration were not passed into law by the Republican Congress of the time. Nevertheless, the Clinton administration did increase funds for alternative energy research, mandate new energy efficiency measures, and enforce emission standards. Additionally, it opened up several areas for oil exploration, including some Alaskan and offshore areas.

Under President George W. Bush

The major energy policy goals of the first two years of Republican president George W. Bush's term were to increase and diversify the sources of America's oil supplies and to make energy security a priority. The Bush administration encouraged efforts to import more Russian crude oil and announced plans to open a new consulate in Equatorial Guinea, an oil-rich nation.

DOMESTIC ENERGY USAGE

Domestic Production

Total domestic energy production has more than doubled since 1949, rising from 31.7 quadrillion Btu (British thermal units) in 1949 to 71.9 quadrillion Btu in 2000. (See Table 1.1 and Figure 1.3.) One quadrillion Btu equals the energy produced by approximately 170 million barrels of crude oil. Large production and consumption figures are given in these units to make it easier to compare the various types of energy, which come in different forms.

TABLE 1.1

Energy production by source, 1949–2000

(quadrillion btu)

Year	Fossil Fuels					Nuclear Electric Power	Hydro-electric Pumped Storage³	Renewable Energy¹						Total
	Coal	Natural Gas (Dry)	Crude Oil²	Natural Gas Plant Liquids	Total			Conventional Hydroelectric Power	Wood, Waste, Alcohol⁴	Geothermal	Solar	Wind	Total	
1949	11.974	5.377	10.683	0.714	28.748	0	(5)	1.425	1.549	0	NA	NA	2.974	31.722
1950	14.060	6.233	11.447	0.823	32.563	0	(5)	1.415	1.562	0	NA	NA	2.978	35.540
1951	14.419	7.416	13.037	0.920	35.792	0	(5)	1.424	1.535	0	NA	NA	2.958	38.751
1952	12.734	7.964	13.281	0.998	34.977	0	(5)	1.466	1.474	0	NA	NA	2.940	37.917
1953	12.278	8.339	13.671	1.062	35.349	0	(5)	1.413	1.419	0	NA	NA	2.831	38.181
1954	10.542	8.682	13.427	1.113	33.764	0	(5)	1.360	1.394	0	NA	NA	2.754	36.518
1955	12.370	9.345	14.410	1.240	37.364	0	(5)	1.360	1.424	0	NA	NA	2.784	40.148
1956	13.306	10.002	15.180	1.283	39.771	0	(5)	1.435	1.416	0	NA	NA	2.851	42.622
1957	13.061	10.605	15.178	1.289	40.133	(s)	(5)	1.516	1.334	0	NA	NA	2.849	42.983
1958	10.783	10.942	14.204	1.287	37.216	0.002	(5)	1.592	1.323	0	NA	NA	2.915	40.133
1959	10.778	11.952	14.933	1.383	39.045	0.002	(5)	1.548	1.353	0	NA	NA	2.901	41.949
1960	10.817	12.656	14.935	1.461	39.869	0.006	(5)	1.608	1.320	0.001	NA	NA	2.929	42.804
1961	10.447	13.105	15.206	1.549	40.307	0.020	(5)	1.656	1.295	0.002	NA	NA	2.953	43.280
1962	10.901	13.717	15.522	1.593	41.732	0.026	(5)	1.816	1.300	0.002	NA	NA	3.119	44.877
1963	11.849	14.513	15.966	1.709	44.037	0.038	(5)	1.771	1.323	0.004	NA	NA	3.098	47.174
1964	12.524	15.298	16.164	1.803	45.789	0.040	(5)	1.886	1.337	0.005	NA	NA	3.228	49.056
1965	13.055	15.775	16.521	1.883	47.235	0.043	(5)	2.059	1.335	0.004	NA	NA	3.398	50.676
1966	13.468	17.011	17.561	1.996	50.035	0.064	(5)	2.062	1.369	0.004	NA	NA	3.435	53.534
1967	13.825	17.943	18.651	2.177	52.597	0.088	(5)	2.347	1.340	0.007	NA	NA	3.694	56.379
1968	13.609	19.068	19.308	2.321	54.306	0.142	(5)	2.349	1.419	0.009	NA	NA	3.778	58.225
1969	13.863	20.446	19.556	2.420	56.286	0.154	(5)	2.648	1.440	0.013	NA	NA	4.102	60.541
1970	14.607	21.666	20.401	2.512	59.186	0.239	(5)	2.634	R1.431	0.011	NA	NA	R4.076	R63.501
1971	13.186	22.280	20.033	2.544	58.042	0.413	(5)	2.824	R1.432	0.012	NA	NA	R4.268	62.723
1972	14.092	22.208	20.041	2.598	58.938	0.584	(5)	2.864	R1.503	0.031	NA	NA	R4.398	63.920
1973	13.992	22.187	19.493	2.569	58.241	0.910	(5)	2.861	R1.529	0.043	NA	NA	R4.433	R63.585
1974	14.074	21.210	18.575	2.471	56.331	1.272	(5)	3.177	R1.540	0.053	NA	NA	R4.769	62.372
1975	14.989	19.640	17.729	2.374	54.733	1.900	(5)	3.155	R1.499	0.070	NA	NA	R4.723	61.357
1976	15.654	19.480	17.262	2.327	54.723	2.111	(5)	2.976	R1.713	0.078	NA	NA	R4.768	61.602
1977	15.755	19.565	17.454	2.327	55.101	2.702	(5)	2.333	R1.838	0.077	NA	NA	R4.249	62.052
1978	14.910	19.485	18.434	2.245	55.074	3.024	(5)	2.937	R2.038	0.064	NA	NA	R5.039	63.137
1979	17.540	20.076	18.104	2.286	58.006	2.776	(5)	2.931	R2.152	0.084	NA	NA	R5.166	R65.948
1980	18.598	19.908	18.249	2.254	59.008	2.739	(5)	2.900	R2.485	0.110	NA	NA	R5.494	67.241
1981	18.377	19.699	18.146	2.307	58.529	3.008	(5)	2.758	2.590	0.123	NA	NA	5.471	67.007
1982	18.639	18.319	18.309	2.191	57.458	3.131	(5)	3.266	2.615	0.105	NA	NA	5.985	66.574
1983	17.247	16.593	18.392	2.184	54.416	3.203	(5)	3.527	2.831	0.129	NA	NA	6.488	64.106
1984	19.719	18.008	18.848	2.274	58.849	3.553	(5)	3.386	2.880	0.165	NA	NA	6.431	68.832
1985	19.325	16.980	18.992	2.241	57.539	4.149	(5)	2.970	R2.864	0.198	(s)	NA	R6.033	R67.720
1986	19.509	16.541	18.376	2.149	56.575	4.471	(5)	3.071	R2.841	0.219	(s)	NA	R6.132	R67.178
1987	20.141	17.136	17.675	2.215	57.167	4.906	(5)	2.635	R2.823	0.229	(s)	NA	R5.687	R67.760
1988	20.738	17.599	17.279	2.260	57.875	5.661	(5)	2.334	R2.937	0.217	(s)	NA	R5.489	69.025
1989	21.346	17.847	16.117	2.158	57.468	5.677	(5)	2.855	R3.060	R0.323	0.059	0.024	R6.322	R69.467
1990	22.456	18.362	15.571	2.175	58.564	6.162	-0.036	3.048	R2.660	R0.343	0.063	0.032	R6.145	R70.835
1991	21.594	18.229	15.701	2.306	57.829	6.580	-0.047	3.021	R2.700	0.348	0.066	0.032	R6.167	R70.528
1992	21.629	18.375	15.223	2.363	57.590	6.608	-0.043	2.617	R2.845	R0.355	R0.067	0.030	R5.915	R70.069
1993	20.249	18.584	14.494	2.408	55.736	6.520	-0.042	2.892	R2.803	R0.369	0.071	0.031	R6.165	R68.378

TABLE 1.1

Energy production by source, 1949–2000 [CONTINUED]

(quadrillion btu)

Year	Fossil Fuels					Nuclear Electric Power	Hydro-electric Pumped Storage³	Renewable Energy¹						Total
	Coal	Natural Gas (Dry)	Crude Oil²	Natural Gas Plant Liquids	Total			Conventional Hydroelectric Power	Wood, Waste, Alcohol⁴	Geothermal	Solar	Wind	Total	
1994	22.111	19.348	14.103	2.391	57.952	6.838	-0.035	2.684	R2.938	R0.364	0.072	0.036	R6.093	R70.848
1995	22.029	19.101	13.887	2.442	57.458	7.177	-0.028	3.207	R3.066	R0.314	0.073	0.035	R6.694	R71.301
1996	22.684	19.363	13.723	2.530	58.299	7.168	-0.032	3.593	R3.126	R0.332	0.075	0.035	R7.160	R72.595
1997	23.211	19.394	13.658	2.495	58.758	6.678	-0.042	3.718	R3.004	R0.322	0.074	R0.033	R7.151	R72.545
1998	R23.935	R19.456	13.235	2.420	R59.047	7.157	-0.046	3.345	R2.976	R0.327	0.074	0.031	R6.752	R72.910
1999	R23.186	R19.126	R12.451	R2.528	R57.291	R7.736	R-0.065	3.305	R3.221	R0.373	R0.073	R0.046	R7.018	R71.980
2000ᵖ	22.663	19.741	12.383	2.607	57.395	8.009	-0.058	2.841	3.275	0.319	0.070	0.051	6.556	71.902

[1] End-use consumption, and electric utility and nonutility electricity net generation.
[2] Includes lease condensate.
[3] Pumped storage facility production minus energy used for pumping.
[4] Alcohol is ethanol blended into motor gasoline.
[5] Included in conventional hydroelectric power.
R=Revised. P=Preliminary. (s)=Less than 0.0005 quadrillion Btu. NA=Not available.
Note: Totals may not equal sum of components due to independent rounding.

SOURCE: "Table 1.2 Energy Production by Source, 1949–2000 (Quadrillion Btu)," in Annual Energy Review 2000, U.S. Department of Energy, Energy Information Administration, Washington, DC, 2001

FIGURE 1.3

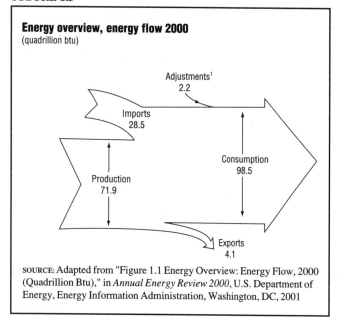

Energy overview, energy flow 2000
(quadrillion btu)

SOURCE: Adapted from "Figure 1.1 Energy Overview: Energy Flow, 2000 (Quadrillion Btu)," in *Annual Energy Review 2000*, U.S. Department of Energy, Energy Information Administration, Washington, DC, 2001

FIGURE 1.4

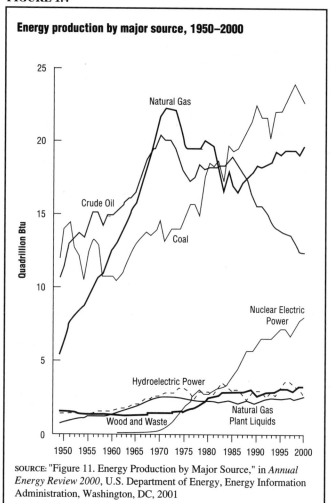

Energy production by major source, 1950–2000

SOURCE: "Figure 11. Energy Production by Major Source," in *Annual Energy Review 2000*, U.S. Department of Energy, Energy Information Administration, Washington, DC, 2001

Table 1.1 and Figure 1.4 show that from 1949 to 2000, the energy produced in the United States from coal generally increased steadily. The energy produced from oil rose until 1972 but by 2000 had declined to about the levels produced in 1950. Likewise, the energy produced from natural gas rose until 1972 and then fell, but it fell only from 1973 through the mid-1980s and has since been on the rise. The energy produced from nuclear power has increased over the past 51 years, while the energy produced from hydroelectric and biofuel power has remained relatively steady. Coal was the leading energy producer in the United States as of 2000, with natural gas in second place. Oil was the third largest source of energy but on the decline, while nuclear electric power, ranked fourth in 2002, was on the rise.

Domestic Consumption

While total domestic energy production has more than doubled since 1949, total domestic energy consumption has more than tripled, rising from 30 quadrillion Btu in 1949 to 98.5 quadrillion Btu in 2000. (See Figure 1.1 and Figure 1.3.) Domestic energy consumption more than doubled from 1949 to 1973, increasing from 30 to 74 quadrillion Btu. Meanwhile, the economy grew at about the same rate, so the increased consumption of energy reflected the growth in the economy—as the nation grew, it used more fuel, mainly more petroleum and natural gas.

However, after the huge 1973 oil price increases, energy consumption fell, rose, and fell again, eventually returning to 1973 levels by 1986. (See Figure 1.1.) Following the drop in crude oil prices in 1986, U.S. imports of oil began to rise, and energy consumption increased, reaching an all-time high of 98.5 quadrillion Btu in 2000. According to U.S. Census Bureau data, the U.S. popula-

tion grew by 88 percent from 1950 to 2000, while energy consumption rose by 181 percent during the same period.

Before the 1973 oil crisis, U.S. energy consumption increased quite quickly. (See Figure 1.1.) After 1973 energy consumption continued to increase but less sharply, as Americans became more efficient and used less energy to accomplish more. Energy consumption shifted slightly away from petroleum and natural gas toward electricity generated by other fuels. In 1973 petroleum and natural gas accounted for 77 percent of total energy consumption; by 2000 their share had dropped to 62.2 percent (38.5 percent petroleum and 23.7 percent natural gas).

Figure 1.5 shows energy production and consumption flows, including types of energy sources, in 2000. Coal, which in 1973 accounted for 17 percent of all energy consumed, accounted for 22.8 percent in 2000, or 22.4 quadrillion Btu out of a total of 98.5 quadrillion Btu. Nuclear power, which contributed barely 1 percent of the nation's consumption in 1973, accounted for 8.1 percent in 2000. Renewable energy sources (hydroelectric, solar, biofuels, and wind energy) accounted for 6.9 percent of energy consumed.

FIGURE 1.5

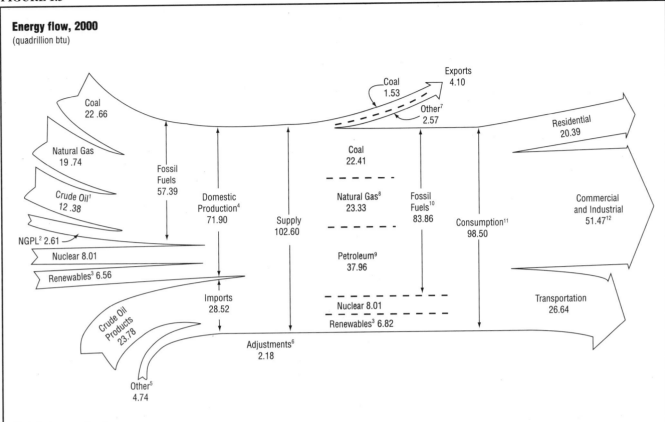

Energy flow, 2000
(quadrillion btu)

[1]Includes lease condensate.
[2]Natural gas plant liquids.
[3]Conventional hydroelectric power, wood, waste, ethanol blended into motor gasoline, geothermal, solar, and wind.
[4]Includes -0.06 quadrillion Btu hydroelectric pumped storage.
[5]Natural gas, coal, coal coke, and electricity.
[6]Stock changes, losses, gains, miscellaneous blending components, and unaccounted-for supply.
[7]Crude oil, petroleum products, natural gas, electricity, and coal coke.
[8]Includes supplemental gaseous fuels.
[9]Petroleum products, including natural gas plant liquids.
[10]Includes 0.07 quadrillion Btu coal coke net imports and 0.10 electricity net imports from fossil fuels.
[11]Includes, in quadrillion Btu, 0.10 electricity net imports from fossil fuels; -0.06 hydroelectric pumped storage; and -0.14 ethanol blended into motor gasoline, which is accounted for in both fossil fuels and renewables and removed once from this total to avoid double-counting.
[12]Commercial and industrial sector totals plus adjustments to avoid double-counting the amount of petroleum, natural gas, and coal that is included under both "End-Use Sectors" and "Electric Power Sector." See Tables 5.12d, 6.5, and 7.3.
Notes: Data are preliminary. Totals may not equal sum of components due to independent rounding.

SOURCE: "Diagram 1. Energy Flow, 2000 (Quadrillion Btu)," in *Annual Energy Review 2000*, U.S. Department of Energy, Energy Information Administration, Washington, DC, 2001

ENERGY IMPORTS AND EXPORTS

After 1958 the United States consumed more energy than it produced but made up the difference by importing energy (see Figure 1.1). Imports (mainly oil) grew rapidly from 1953 through 1973 as the United States grew its economy using inexpensive oil. In 1973 net imports of petroleum reached almost 13 quadrillion Btu. (See Figure 1.6.)

Although the Arab oil embargo of 1973–74, coupled with increased oil prices, momentarily slowed growth in petroleum imports, the general increase continued, with imports exceeding 18 quadrillion Btu in 1977. That year, U.S. dependence on petroleum imports rose to 46.5 percent of the nation's oil consumption. Despite the lesson of 1973, it took a second round of price increases in 1979–80, as shown in Figure 1.2, accompanied by lengthy

and frustrating lines at gas stations, to convince Americans that they had to become less dependent on imported oil, conserve resources more, or both. Oil imports declined in 1985, and U.S. dependence on foreign oil decreased sharply to 27.3 percent of oil consumption. (See Figure 1.7.)

Nonetheless, when the price of crude oil dropped again, the demand returned, and U.S. dependence on foreign sources of oil increased. When Iraq invaded Kuwait in 1990, this potential threat to the flow of oil to America and other industrialized nations was one of the reasons the United States challenged Saddam Hussein and eventually declared war. By 2000 imported oil accounted for a record 51.6 percent of U.S. oil consumption. (See Figure 1.7.) In 2000 net imports of petroleum reached slightly over 20

FIGURE 1.6

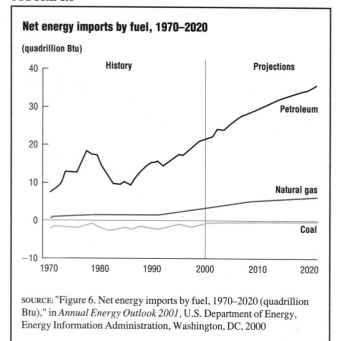

Net energy imports by fuel, 1970–2020

SOURCE: "Figure 6. Net energy imports by fuel, 1970–2020 (quadrillion Btu)," in *Annual Energy Outlook 2001*, U.S. Department of Energy, Energy Information Administration, Washington, DC, 2000

FIGURE 1.7

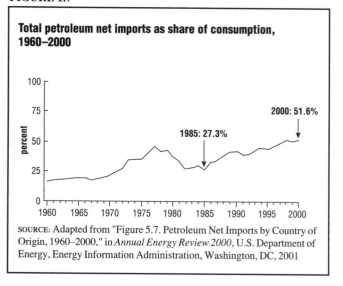

Total petroleum net imports as share of consumption, 1960–2000

SOURCE: Adapted from "Figure 5.7. Petroleum Net Imports by Country of Origin, 1960–2000," in *Annual Energy Review 2000*, U.S. Department of Energy, Energy Information Administration, Washington, DC, 2001

FIGURE 1.8

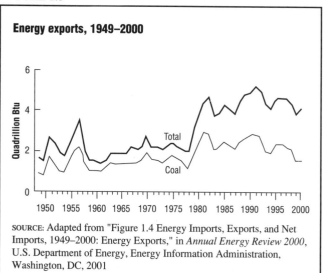

Energy exports, 1949–2000

SOURCE: Adapted from "Figure 1.4 Energy Imports, Exports, and Net Imports, 1949–2000: Energy Exports," in *Annual Energy Review 2000*, U.S. Department of Energy, Energy Information Administration, Washington, DC, 2001

quadrillion Btu. (See Figure 1.6.) However, after the terrorist events of September 11, 2001, and the subsequent Bush administration "War on Terror," the concept of energy independence, or at least less energy dependence, became increasingly important.

Although the United States imports energy in the form of oil, it exports energy in the form of coal. Since 1950 America has produced more coal than it has consumed and has been an exporter of coal to other nations. In 2000 coal exports totaled 1.53 quadrillion Btu, nearly 40 percent of U.S. energy exports. (See Figure 1.8 and Table 1.2.)

FOSSIL FUEL PRODUCTION PRICES

Production prices are the value of fuel produced. The combined production prices of fossil fuels (crude oil, natural gas, and coal) slowly declined from 1949 through 1972. (See Figure 1.9 and Table 1.3.) These prices then increased dramatically from 1973 through 1981, and fell through 1998. To indicate how marked this decline in fossil fuel prices was, the composite value of all fossil fuel prices (in real dollars, which account for inflation) dropped by over two-thirds from 1981 to 1998, from $4.40 per million Btu to $1.36 per million Btu. These huge drops meant economic problems in fuel-producing American states such as Texas, Louisiana, Oklahoma, Montana, West Virginia, and Ohio, and in energy-exporting nations such as many Middle Eastern nations, Nigeria, Indonesia, Venezuela, and Trinidad. On the other hand, they were a windfall for industries that used a lot of energy, such as airlines, trucking companies, steel mills, and electric utilities. Since 1998 fossil fuel production prices have been climbing, reaching $2.40 per million Btu in 2000.

The production prices of both crude oil, the most expensive of the fossil fuels, and natural gas followed a pattern of rising and falling similar to that of the fossil fuel composite price from 1949 to 2000. (See Figure 1.9 and Table 1.3.) After slowly declining from 1949 through 1972, crude oil production prices rose the most dramatically of all the fossil fuels from 1973 to 1981, topping out at $8.78 per million Btu in 1981. The price then tumbled to $1.82 in 1998. However, it then rose sharply during the next two years, reaching $4.31 in 2000. For natural gas, the price sank from $3.37 per million Btu in 1983 to $1.69 in 1998. It jumped back up to $3.03 in 2000.

The story of coal prices is a little different from that of the other fossil fuels. Coal production prices rose from 1970 to 1975, but then declined steadily through 2000. (See Figure 1.9 and Table 1.3.) Its peak price in 1975 was $2.11 per million Btu in real dollars, and in 2000 it dropped to a low of 74 cents per million Btu.

TABLE 1.2

Energy imports, exports, and net imports, 1949–2000

(quadrillion btu)

Year	Imports					Exports					Net Imports				
	Coal	Natural Gas	Petroleum[1]	Other[2]	Total	Coal	Natural Gas	Petroleum	Other[2]	Total	Coal	Natural Gas	Petroleum[1]	Other[2]	Total
1949	0.01	0.00	1.43	0.03	1.47	0.88	0.02	0.68	0.02	1.59	-0.87	-0.02	0.75	0.02	-0.13
1950	0.01	0.00	1.89	0.04	1.93	0.79	0.03	0.64	0.01	1.47	-0.78	-0.03	1.24	0.03	0.47
1951	0.01	0.00	1.87	0.04	1.92	1.68	0.03	0.89	0.03	2.62	-1.67	-0.03	0.98	0.01	-0.71
1952	0.01	0.01	2.11	0.04	2.17	1.40	0.03	0.91	0.02	2.37	-1.40	-0.02	1.20	0.02	-0.20
1953	0.01	0.01	2.28	0.04	2.34	0.98	0.03	0.84	0.01	1.87	-0.97	-0.02	1.44	0.02	0.47
1954	0.01	0.01	2.32	0.04	2.37	0.91	0.03	0.75	0.02	1.70	-0.91	-0.02	1.58	0.04	0.67
1955	0.01	0.01	2.75	0.06	2.83	1.46	0.03	0.77	0.02	2.29	-1.46	-0.02	1.98	0.04	0.54
1956	0.01	0.01	3.17	0.06	3.25	1.98	0.04	0.91	0.02	2.95	-1.98	-0.02	2.26	0.02	0.30
1957	0.01	0.04	3.46	0.06	3.57	2.17	0.04	1.20	0.03	3.45	-2.16	(s)	2.26	0.03	0.12
1958	0.01	0.14	3.72	0.05	3.92	1.42	0.04	0.58	0.02	2.06	-1.41	0.10	3.14	0.03	1.86
1959	0.01	0.14	3.91	0.05	4.11	1.05	0.02	0.45	0.02	1.54	-1.04	0.12	3.46	0.04	2.57
1960	0.01	0.16	4.00	0.06	4.23	1.02	0.01	0.43	0.02	1.48	-1.02	0.15	3.57	0.02	2.74
1961	(s)	0.23	4.19	0.04	4.46	0.98	0.01	0.37	0.02	1.38	-0.98	0.22	3.82	(s)	3.08
1962	0.01	0.42	4.56	0.03	5.01	1.08	0.02	0.36	0.03	1.48	-1.08	0.40	4.20	-0.01	3.53
1963	0.01	0.42	4.65	0.03	5.10	1.36	0.02	0.44	0.03	1.85	-1.35	0.40	4.21	0.01	3.25
1964	0.01	0.46	4.96	0.07	5.49	1.34	0.02	0.43	0.06	1.84	-1.33	0.44	4.53	-0.02	3.65
1965	(s)	0.47	5.40	0.04	5.92	1.38	0.03	0.39	0.06	1.85	-1.37	0.44	5.01	-0.01	4.06
1966	(s)	0.50	5.63	0.05	6.18	1.35	0.03	0.41	0.06	1.85	-1.35	0.47	5.21	-0.02	4.32
1967	0.01	0.58	5.56	0.04	6.19	1.35	0.08	0.65	0.06	2.15	-1.37	0.50	4.91	-0.02	4.04
1968	0.01	0.67	6.21	0.04	6.93	1.38	0.10	0.49	0.06	2.03	-1.37	0.58	5.73	-0.02	4.90
1969	(s)	0.75	6.90	0.06	7.71	1.53	0.05	0.49	0.06	2.15	-1.53	0.70	6.42	-0.02	5.56
1970	(s)	0.85	7.47	0.07	8.39	1.94	0.07	0.55	0.11	2.66	-1.93	0.77	6.92	-0.04	5.72
1971	(s)	0.96	8.54	0.08	9.58	1.55	0.08	0.47	0.07	2.18	-1.54	0.88	8.07	(s)	7.41
1972	(s)	1.05	10.30	0.11	11.46	1.53	0.08	0.47	0.06	2.14	-1.53	0.97	9.83	0.05	9.32
1973	(s)	1.06	13.47	0.20	14.73	1.43	0.08	0.49	0.06	2.05	-1.42	0.98	12.98	0.14	12.68
1974	0.05	0.99	13.13	0.25	14.41	1.62	0.08	0.46	0.08	2.22	-1.57	0.91	12.66	0.19	12.19
1975	0.02	0.98	12.95	0.16	14.11	1.76	0.07	0.44	0.06	2.36	-1.74	0.90	12.51	0.08	11.75
1976	0.03	0.99	15.67	0.15	16.84	1.60	0.07	0.47	0.06	2.19	-1.57	0.92	15.20	0.09	14.65
1977	0.04	1.04	18.76	0.26	20.09	1.44	0.06	0.51	0.06	2.07	-1.40	0.98	18.24	0.20	18.02
1978	0.07	0.99	17.82	0.36	19.25	1.08	0.05	0.77	0.03	1.93	-1.00	0.94	17.06	0.33	17.32
1979	0.05	1.30	17.93	0.33	19.62	1.75	0.06	1.00	0.06	2.87	-1.70	1.24	16.93	0.27	16.75
1980	0.03	1.01	14.66	0.28	15.97	2.42	0.05	1.16	0.09	3.72	-2.39	0.96	13.50	0.18	12.25
1981	0.03	0.92	12.64	0.39	13.97	2.94	0.06	1.26	0.06	4.33	-2.92	0.86	11.38	0.33	9.65
1982	0.02	0.95	10.78	0.35	12.09	2.79	0.05	1.73	0.06	4.63	-2.77	0.90	9.05	0.28	7.46
1983	0.03	0.94	10.65	0.41	12.03	2.04	0.06	1.57	0.05	3.72	-2.01	0.90	9.08	0.36	8.31
1984	0.03	0.85	11.43	0.46	12.77	2.15	0.06	1.54	0.05	3.80	-2.12	0.79	9.89	0.40	8.96
1985	0.05	0.95	10.61	0.49	12.10	2.44	0.06	1.66	0.08	4.23	-2.39	0.90	8.95	0.41	7.87
1986	0.06	0.75	13.20	0.43	14.44	2.25	0.06	1.67	0.08	4.06	-2.19	0.69	11.53	0.36	10.38
1987	0.04	0.99	14.16	0.57	15.76	2.09	0.05	1.63	0.08	3.85	-2.05	0.94	12.53	0.49	11.91
1988	0.05	1.30	15.75	0.47	17.56	2.50	0.07	1.74	0.10	4.42	-2.45	1.22	14.01	0.37	13.15
1989	0.07	1.39	17.16	0.34	18.96	2.64	0.11	1.84	0.18	4.77	-2.57	1.28	15.33	0.15	14.19
1990	0.07	1.55	17.12	0.22	18.95	2.77	0.09	1.82	0.04	4.87	-2.70	1.46	15.29	0.03	14.09
1991	0.08	1.80	16.35	0.27	18.50	2.85	0.13	2.13	0.05	5.16	-2.77	1.67	14.22	0.22	13.34
1992	0.10	2.16	16.97	0.35	19.58	2.68	0.22	2.01	0.05	4.96	-2.59	1.94	14.96	0.31	14.62
1993	0.20	2.40	18.51	0.39	21.50	1.96	0.14	2.12	0.06	4.28	-1.76	2.25	16.40	0.32	17.22
1994	0.22	2.68	19.24	0.58	22.73	1.88	0.16	1.99	0.05	4.08	-1.66	2.52	17.26	0.53	18.65

TABLE 1.2

Energy imports, exports, and net imports, 1949–2000 [CONTINUED]

(quadrillion btu)

Year	Imports						Exports						Net Imports				
	Coal	Natural Gas	Petroleum[1]	Other[2]	Total		Coal	Natural Gas	Petroleum	Other[2]	Total		Coal	Natural Gas	Petroleum[1]	Other[2]	Total
1995	0.24	2.90	R18.88	0.55	R22.57		2.32	0.16	1.99	0.07	4.54		-2.08	2.74	R16.89	0.47	R18.03
1996	0.20	3.00	R20.29	0.52	R24.01		2.37	0.16	2.06	0.07	4.66		-2.17	2.85	R18.23	0.45	R19.35
1997	0.19	3.06	21.74	0.52	R25.51		2.19	0.16	2.10	0.12	R4.58		-2.01	2.90	19.64	0.40	20.94
1998	0.22	3.22	22.91	0.50	26.86		R2.09	0.16	1.97	0.16	R4.39		R-1.87	3.06	20.94	0.34	R22.47
1999	0.23	R3.66	R23.13	0.52	R27.55		1.53	0.16	R1.95	0.17	R3.81		R-1.30	R3.50	R21.18	0.36	R23.74
2000P	0.31	3.81	23.78	0.61	28.52		1.53	0.24	2.15	0.18	4.10		-1.21	3.57	21.63	0.43	24.42

[1] Includes imports into the Strategic Petroleum Reserve, which began in 1977.
[2] Coal coke and small amounts of electricity transmitted across U.S. borders with Canada and Mexico.
R=Revised. P=Preliminary. (s)=Less than 0.005 quadrillion Btu and greater than -0.005 quadrillion Btu.
Notes: Includes trade between the United States (50 States and the District of Columbia) and its territories and possessions. · Totals or net import items may not equal sum of components due to independent rounding.

SOURCE: "Table 1.4 Energy Imports, Exports, and Net Imports, 1949–2000 (Quadrillion Btu)," in *Annual Energy Review 2000*, U.S. Department of Energy, Energy Information Administration, Washington, DC, 2001

FIGURE 1.9

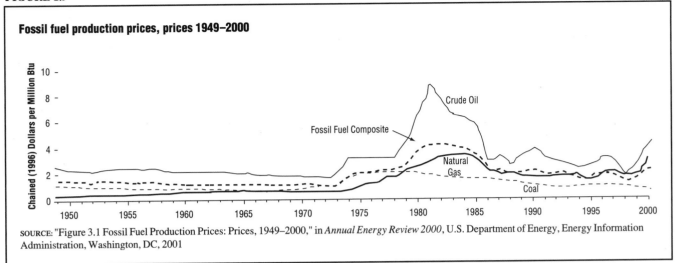

Fossil fuel production prices, prices 1949–2000

SOURCE: "Figure 3.1 Fossil Fuel Production Prices: Prices, 1949–2000," in *Annual Energy Review 2000*, U.S. Department of Energy, Energy Information Administration, Washington, DC, 2001

FIGURE 1.10

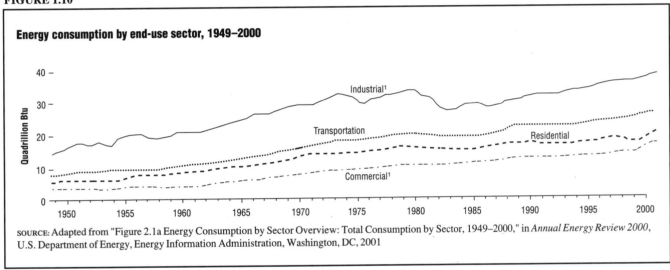

Energy consumption by end-use sector, 1949–2000

SOURCE: Adapted from "Figure 2.1a Energy Consumption by Sector Overview: Total Consumption by Sector, 1949–2000," in *Annual Energy Review 2000*, U.S. Department of Energy, Energy Information Administration, Washington, DC, 2001

ENERGY USE BY SECTOR

Energy use can be classified into four main "end-use" sectors: residential, commercial, industrial, and transportation. Historically, industry has been the largest energy-consuming sector of the economy, followed by the transportation, residential, and commercial sectors, respectively. In 2000 industry used about 39 quadrillion Btu, compared to approximately 27 quadrillion Btu in the transportation sector, 20 quadrillion Btu in the residential sector, and 17 quadrillion Btu in the commercial sector. (See Figure 1.10.)

Within sectors, energy sources have changed over time. For example, in the commercial sector, coal was the leading energy source through 1953 but declined dramatically in favor of petroleum (through 1963) and then natural gas (since 1963). (See Table 1.4.) Similarly, coal was the leading energy source in the residential sector through 1949. (See Table 1.5.) Natural gas quickly took over, with petroleum in second place. Industry used more coal than natural gas or petroleum through 1957, but after that year, natural gas and petroleum took over as nearly equally preferred energy sources. (See Table 1.6.) In transportation, reliance on petroleum has been increasing since 1949. (See Table 1.7.)

Not included in the four main sectors of energy consumption is the electric power sector. This sector includes electric utilities that generate, transmit, distribute, and sell electricity for use by the public. Electricity use in the four main sectors of energy consumption is noted in Table 1.4, Table 1.5, Table 1.6, and Table 1.7. To generate electricity that comes into homes, businesses, and industry, the electricity sector must harness energy from renewable sources such as the sun, the wind, water movement, and geothermal sources, or it must generate electricity from nuclear power or the burning of fossil fuels. Table 1.8 shows that the electric power sector generates most of the electricity for the United States by burning coal. The electric power sector consumed 20,503 trillion Btu of coal in 2000, a figure that has increased tenfold since 1949. Nuclear power is the next most utilized fuel, followed by natural gas, hydroelectric (water power), and petroleum. Renewable sources

TABLE 1.3

Fossil fuel production prices, 1949–2000

(dollars per million btu)

Year	Coal[1] Nominal	Coal[1] Real[5]	Natural Gas[2] Nominal	Natural Gas[2] Real[5]	Crude Oil[3] Nominal	Crude Oil[3] Real[5]	Fossil Fuel Composite[4] Nominal	Fossil Fuel Composite[4] Real[5]	Percent Change[6]
1949	0.21	1.22	0.05	0.31	0.44	2.54	0.26	1.52	—
1950	0.21	1.19	0.06	0.36	0.43	2.48	0.26	1.46	-3.8
1951	0.21	1.13	0.06	0.34	0.44	2.33	0.26	1.38	-5.3
1952	0.21	1.10	0.07	0.38	0.44	2.30	0.26	1.37	-0.7
1953	0.21	1.08	0.08	0.42	0.46	2.40	0.27	1.42	3.2
1954	0.19	0.99	0.09	0.46	0.48	2.46	0.28	1.42	0.5
1955	0.19	0.94	0.09	0.45	0.48	2.42	0.27	1.37	-3.9
1956	0.20	0.97	0.10	0.48	0.48	2.35	0.28	1.36	-0.8
1957	0.21	0.99	0.10	0.47	0.53	2.52	0.30	1.42	4.1
1958	0.20	0.94	0.11	0.50	0.52	2.40	0.29	1.35	-4.7
1959	0.20	0.91	0.12	0.54	0.50	2.28	0.29	1.31	-3.1
1960	0.19	0.87	0.13	0.57	0.50	2.24	0.28	1.28	-2.4
1961	0.19	0.85	0.14	0.60	0.50	2.22	0.29	1.28	0.0
1962	0.19	0.82	0.14	0.64	0.50	2.20	0.29	1.27	-0.7
1963	0.18	0.80	0.14	0.63	0.50	2.16	0.28	1.23	-2.8
1964	0.18	0.79	0.14	0.58	0.50	2.13	0.28	1.19	-3.5
1965	0.18	0.77	0.14	0.61	0.49	2.07	0.28	1.16	-1.9
1966	0.19	0.77	0.14	0.59	0.50	2.03	0.28	1.14	-1.7
1967	0.19	0.76	0.14	0.58	0.50	2.00	0.28	1.13	-1.6
1968	0.19	0.74	0.14	0.54	0.51	1.93	0.28	1.08	-3.8
1969	0.21	0.76	0.15	0.56	0.53	1.93	0.30	1.08	0.0
1970	0.27	0.92	0.15	0.53	0.55	1.89	0.32	1.09	0.9
1971	0.30	1.00	0.16	0.53	0.58	1.91	0.34	1.11	1.8
1972	0.33	1.04	0.17	0.54	0.58	1.84	0.35	1.10	-1.3
1973	0.37	1.09	0.20	0.60	0.67	2.00	0.40	1.18	7.7
1974	0.69	1.87	0.27	0.74	1.18	3.23	0.68	1.85	55.8
1975	0.84	2.11	0.40	1.00	1.32	3.30	0.82	2.05	11.1
1976	0.86	2.02	0.53	1.26	1.41	3.34	0.90	2.13	3.9
1977	0.88	1.96	0.72	1.61	1.48	3.28	1.01	2.24	5.0
1978	0.98	2.04	0.84	1.73	1.55	3.22	1.12	2.31	3.3
1979	1.06	2.02	1.08	2.07	2.18	4.17	1.42	2.71	17.3
1980	1.10	1.93	1.45	2.54	3.72	6.52	2.04	3.58	32.0
1981	1.18	1.90	1.80	2.88	5.48	8.78	2.74	4.40	22.9
1982	1.22	1.85	2.22	3.35	4.92	7.42	2.76	4.16	-5.4
1983	1.18	1.71	2.32	3.37	4.52	6.56	2.70	3.92	-5.8
1984	1.16	1.63	2.40	3.36	4.46	6.25	2.65	3.70	-5.5
1985	1.15	1.56	2.26	3.06	4.15	5.64	2.51	3.41	-8.0
1986	1.09	1.44	1.75	2.32	2.16	2.86	1.65	2.20	-35.6
1987	1.05	1.36	1.50	1.94	2.66	3.42	1.70	2.19	-0.2
1988	1.01	1.26	1.52	1.90	2.17	2.70	1.53	1.91	-12.8
1989	1.00	1.20	1.53	1.83	2.73	3.28	1.67	2.01	5.0
1990	1.00	1.15	1.55	1.79	3.45	3.99	1.84	2.13	6.1
1991	0.99	1.10	1.48	1.65	2.85	3.18	1.67	1.86	-12.5
1992	0.97	1.06	1.57	1.71	2.76	3.00	1.66	1.80	-3.1
1993	0.93	0.99	1.84	1.96	2.46	2.61	1.67	1.78	-1.6
1994	0.91	0.94	1.67	1.74	2.27	2.37	1.53	1.59	-10.5
1995	0.88	0.90	1.40	1.43	2.52	2.57	1.47	1.50	-5.5
1996	0.87	0.87	1.96	1.96	3.18	3.18	1.82	1.82	21.3
1997	0.85	0.84	2.10	2.06	2.97	R2.91	1.81	1.77	R-2.7
1998	0.82	R0.80	1.75	R1.69	1.87	1.82	R1.40	R1.36	R-23.3
1999	R0.80	R0.76	R1.95	R1.86	2.68	R2.56	R1.64	R1.57	R15.5
2000P	70.80	70.74	3.24	3.03	4.61	4.31	72.57	72.40	753.2

[1]Bituminous coal, subbituminous coal, and lignite prices are based on the value of coal produced at free-on-board (f.o.b.) mines; anthracite prices through 1978 are f.o.b. preparation plants and for 1979 forward are f.o.b. mines.
[2]Wellhead prices.
[3]Domestic first purchase prices.
[4]Derived by multiplying the price per Btu of each fossil fuel by the total Btu content of the production of each fossil fuel and dividing this accumulated value of total fossil fuel production by the accumulated Btu content of total fossil fuel production.
[5]In chained (1996) dollars, calculated by using gross domestic product implicit price deflators.
[6]Based on real values.
[7]Calculated using the 1999 coal price for the 2000 value.
R=Revised. P=Preliminary. — = Not applicable.

SOURCE: "Table 3.1 Fossil Fuel Production Prices, 1949–2000 (Dollars per Million Btu)," in *Annual Energy Review 2000*, U.S. Department of Energy, Energy Information Administration, Washington, DC, 2001

TABLE 1.4

Commercial sector energy consumption, 1949–2000

(trillion btu)

Year	Coal[1]	Natural Gas[1,2]	Petroleum[1]	Wood[3]	Geothermal[4]	Electricity[5]
	Fossil Fuels			**Renewable Energy**		
1949	1,554	360	727	20	NA	200
1950	1,542	401	862	19	NA	225
1951	1,331	481	924	18	NA	252
1952	1,169	534	942	17	NA	273
1953	985	549	970	16	NA	297
1954	825	605	1,000	15	NA	319
1955	801	651	1,081	15	NA	350
1956	730	742	1,122	14	NA	380
1957	535	803	1,083	13	NA	411
1958	512	902	1,125	13	NA	435
1959	415	1,009	1,194	12	NA	488
1960	407	1,056	1,228	12	NA	543
1961	371	1,115	1,247	11	NA	572
1962	371	1,249	1,280	11	NA	621
1963	317	1,307	1,262	10	NA	688
1964	274	1,419	1,247	9	NA	738
1965	259	1,490	1,386	9	NA	789
1966	263	1,676	1,436	9	NA	859
1967	225	2,022	1,483	8	NA	925
1968	203	2,140	1,510	8	NA	1,014
1969	195	2,323	1,520	8	NA	1,108
1970	165	2,473	1,551	8	NA	1,201
1971	175	2,587	1,510	7	NA	1,288
1972	153	2,678	1,530	7	NA	1,408
1973	160	2,649	1,565	7	NA	1,517
1974	175	2,617	1,423	7	NA	1,501
1975	147	2,558	1,310	8	NA	1,598
1976	144	2,718	1,461	9	NA	1,678
1977	148	2,548	1,511	10	NA	1,754
1978	165	2,643	1,450	12	NA	1,813
1979	149	2,836	1,334	14	NA	1,854
1980	115	2,674	1,287	21	NA	1,906
1981	137	2,583	1,090	21	NA	2,033
1982	155	2,673	1,008	22	NA	2,077
1983	162	2,508	1,136	22	NA	2,116
1984	171	2,600	1,198	22	NA	2,264
1985	141	2,508	1,039	24	NA	2,351
1986	141	2,386	1,099	27	NA	2,439
1987	129	2,505	1,079	29	NA	2,539
1988	136	2,748	1,037	32	NA	2,675
1989	118	2,802	966	34	3	2,767
1990	129	2,701	907	37	3	2,860
1991	118	2,813	861	39	3	2,918
1992	118	2,890	813	42	3	2,900
1993	119	2,942	753	44	3	3,019
1994	118	2,979	753	45	4	3,116
1995	117	3,113	715	45	5	3,252
1996	122	3,244	747	49	5	3,344
1997	129	3,302	709	47	6	3,503
1998	92	3,098	665	47	7	3,678
1999	103	3,130	672	R51	7	3,766
2000P	102	3,452	696	52	8	3,867

[1]Includes some consumption at nonutilities.
[2]Includes supplemental gaseous fuels.
[3]Wood only.
[4]Geothermal heat pump and direct use energy.
[5]Electric utility retail sales of electricity, including nonutility sales of electricity to utilities for distribution to end users; beginning in 1996, also includes sales to ultimate consumers by power marketers.
R=Revised. P=Preliminary. NA=Not available.

SOURCE: "Table 2.1c Commercial Sector Energy Consumption, 1949–2000 (Trillion Btu)," in *Annual Energy Review 2000*, U.S. Department of Energy, Energy Information Administration, Washington, DC, 2001

TABLE 1.5

Residential sector energy consumption, 1949–2000

(trillion btu)

Year	Primary Consumption Fossil Fuels Coal	Natural Gas[1]	Petroleum	Renewable Energy Wood[2]	Geothermal[3]	Solar[4]	Electricity[5]
1949	1,272	1,027	1,121	1,055	NA	NA	228
1950	1,261	1,240	1,340	1,006	NA	NA	246
1951	1,134	1,526	1,481	958	NA	NA	284
1952	1,079	1,679	1,522	899	NA	NA	319
1953	946	1,744	1,533	832	NA	NA	355
1954	858	1,961	1,667	800	NA	NA	397
1955	867	2,198	1,792	775	NA	NA	438
1956	823	2,409	1,880	739	NA	NA	490
1957	654	2,588	1,828	702	NA	NA	535
1958	652	2,809	1,994	688	NA	NA	578
1959	573	3,015	1,989	647	NA	NA	630
1960	585	3,212	2,265	627	NA	NA	687
1961	534	3,362	2,332	587	NA	NA	732
1962	512	3,600	2,441	560	NA	NA	794
1963	438	3,700	2,459	537	NA	NA	856
1964	379	3,908	2,375	499	NA	NA	928
1965	358	4,028	2,481	468	NA	NA	993
1966	349	4,275	2,471	455	NA	NA	1,081
1967	299	4,451	2,557	434	NA	NA	1,160
1968	269	4,588	2,685	426	NA	NA	1,302
1969	248	4,875	2,739	415	NA	NA	1,456
1970	209	4,987	2,755	401	NA	NA	1,591
1971	175	5,126	2,777	382	NA	NA	1,704
1972	116	5,264	2,895	380	NA	NA	1,838
1973	94	4,977	2,825	354	NA	NA	1,976
1974	82	4,901	2,573	371	NA	NA	1,973
1975	63	5,023	2,495	425	NA	NA	2,007
1976	59	5,147	2,720	482	NA	NA	2,069
1977	57	4,913	2,695	542	NA	NA	2,202
1978	49	4,981	2,620	622	NA	NA	2,301
1979	37	5,055	2,114	728	NA	NA	2,330
1980	31	4,866	1,748	R859	NA	NA	2,448
1981	30	4,660	1,543	869	NA	NA	2,464
1982	32	4,753	1,441	937	NA	NA	2,489
1983	31	4,516	1,362	925	NA	NA	2,562
1984	38	4,692	1,337	923	NA	NA	2,662
1985	35	4,571	1,483	899	NA	NA	2,709
1986	35	4,439	1,457	876	NA	NA	2,795
1987	32	4,449	1,508	852	NA	NA	2,902
1988	32	4,765	1,563	885	NA	NA	3,046
1989	28	4,929	1,560	918	R5	53	3,090
1990	26	4,523	1,266	581	R6	56	3,153
1991	23	4,697	1,293	613	R6	58	3,260
1992	24	4,835	1,312	645	R6	60	3,193
1993	24	5,095	1,387	548	R7	62	3,394
1994	21	4,988	1,340	537	R6	64	3,441
1995	17	4,981	1,361	596	R7	65	3,557
1996	17	5,383	1,492	595	R7	66	3,694
1997	16	5,118	1,454	433	R7	65	3,671
1998	13	4,669	1,324	R387	R8	65	3,856
1999	14	4,858	1,456	R414	R8	R64	3,906
2000P	14	5,061	1,475	433	9	62	4,066

[1]Includes supplemental gaseous fuels.
[2]Wood only.
[3]Geothermal heat pump and direct use energy.
[4]Solar thermal direct use and photovoltaic energy. Includes small amounts of commercial sector use.
[5]Electric utility retail sales of electricity, including nonutility sales of electricity to utilities for distribution to end users; beginning in 1996, also includes sales to ultimate consumers by power marketers.
R=Revised. P=Preliminary. NA=Not available.

SOURCE: "Table 2.1b Residential Sector Energy Consumption, 1949–2000 (Trillion Btu)," in *Annual Energy Review 2000*, U.S. Department of Energy, Energy Information Administration, Washington, DC, 2001

Energy: Shortage, Glut, or Enough?

TABLE 1.6

Industrial sector energy consumption, 1949–2000

(trillion btu)

Year	Primary Consumption						Electricity[4]
	Fossil Fuels			Renewable Energy			
	Coal[1]	Natural Gas[1,2]	Petroleum[1]	Wood	Waste	Geothermal[3]	
1949	5,433	3,188	3,469	468	NA	NA	418
1950	5,781	3,546	13,279	532	NA	NA	500
1951	6,202	4,052	14,503	553	NA	NA	567
1952	5,517	4,181	14,047	552	NA	NA	601
1953	5,931	4,304	14,707	566	NA	NA	678
1954	4,730	4,319	13,674	576	NA	NA	711
1955	5,620	4,701	15,421	631	NA	NA	887
1956	5,667	4,874	15,865	661	NA	NA	976
1957	5,536	5,107	15,863	616	NA	NA	1,003
1958	4,533	5,208	15,142	620	NA	NA	978
1959	4,413	5,647	15,791	692	NA	NA	1,075
1960	4,543	5,973	16,259	680	NA	NA	1,107
1961	4,345	6,170	16,261	695	NA	NA	1,149
1962	4,385	6,451	16,826	728	NA	NA	1,228
1963	4,590	6,748	17,557	775	NA	NA	1,288
1964	4,915	7,114	18,564	827	NA	NA	1,382
1965	5,127	7,339	19,236	855	NA	NA	1,463
1966	5,215	7,795	20,094	902	NA	NA	1,582
1967	4,934	8,043	20,081	895	NA	NA	1,655
1968	4,855	8,626	20,853	982	NA	NA	1,778
1969	4,712	9,234	21,606	1,014	NA	NA	1,909
1970	4,656	9,536	21,923	1,019	NA	NA	1,948
1971	3,944	9,892	21,661	1,040	NA	NA	2,011
1972	3,993	9,884	22,386	1,113	NA	NA	2,187
1973	4,057	10,388	23,539	1,165	NA	NA	2,341
1974	3,870	10,004	22,624	1,159	NA	NA	2,337
1975	3,667	8,532	20,360	1,063	NA	NA	2,346
1976	3,661	8,762	21,436	1,220	NA	NA	2,573
1977	3,454	8,635	21,880	1,281	NA	NA	2,682
1978	3,314	8,539	21,843	1,400	NA	NA	2,761
1979	3,593	8,549	22,771	1,405	NA	NA	2,873
1980	3,155	8,395	21,043	1,600	NA	NA	2,781
1981	3,157	8,257	19,684	1,602	87	NA	2,817
1982	2,552	7,121	17,446	1,516	118	NA	2,542
1983	2,490	6,826	16,718	1,690	155	NA	2,648
1984	2,842	7,448	18,293	1,679	204	NA	R2,859
1985	2,760	7,080	17,634	1,645	230	NA	2,855
1986	2,641	6,690	17,235	1,610	256	NA	2,834
1987	2,673	7,323	18,154	1,576	282	NA	2,928
1988	2,828	7,696	18,995	1,625	308	NA	3,059
1989	2,787	8,131	19,078	R1,394	250	R2	R3,501
1990	2,756	8,502	19,582	R1,254	271	R2	3,582
1991	2,601	8,619	19,288	R1,190	275	R2	3,609
1992	2,515	8,967	20,152	R1,233	289	R2	3,734
1993	2,496	9,410	20,383	R1,255	288	R2	3,767
1994	2,510	9,560	20,977	R1,342	318	R3	3,920
1995	2,488	10,064	R21,236	R1,402	322	R3	3,964
1996	R2,434	10,393	R21,912	R1,441	363	R3	4,035
1997	R2,395	10,307	R22,066	R1,513	338	R3	4,051
1998	R2,335	10,168	R21,675	R1,564	312	R3	4,132
1999	R2,243	R10,360	R9,394	R1,711	291	R4	4,255
2000[P]	2,280	10,943	9,197	1,702	287	4	4,355

[1]Includes some consumption at nonutilities.
[2]Includes supplemental gaseous fuels.
[3]Geothermal heat pump and direct use energy.
[4]Electric utility retail sales of electricity, including nonutility sales of electricity to utilities for distribution to end users; beginning in 1989, also includes nonutility facility use of onsite net electricity generation, and electricity sold by nonutilities directly to end users; beginning in 1996, also includes sales to ultimate consumers by power marketers.
R=Revised. P=Preliminary. NA=Not available. (s)=Less than +0.5 trillion Btu and greater than -0.5 trillion Btu.

SOURCE: "Table 2.1d Industrial Sector Energy Consumption, 1949–2000 (Trillion Btu)," in *Annual Energy Review 2000,* U.S. Department of Energy, Energy Information Administration, Washington, DC, 2001

TABLE 1.7

Transportation sector energy consumption, 1949–2000

(trillion btu)

| Year | Primary Consumption | | | Renewable | |
| | Fossil Fuels | | | | |
	Coal	Natural Gas[1]	Petroleum	Alcohol Fuels[2]	Electricity[3]
1949	1,727	NA	6,152	NA	22
1950	1,564	130	6,690	NA	23
1951	1,379	199	7,356	NA	24
1952	984	214	7,709	NA	22
1953	733	238	8,060	NA	22
1954	461	239	8,123	NA	20
1955	421	254	8,801	NA	20
1956	340	306	9,145	NA	19
1957	241	310	9,286	NA	16
1958	115	323	9,514	NA	15
1959	88	362	9,849	NA	14
1960	75	359	10,127	NA	10
1961	19	391	10,324	NA	10
1962	17	396	10,774	NA	10
1963	16	437	11,167	NA	10
1964	17	450	11,497	NA	10
1965	16	517	11,867	NA	10
1966	15	553	12,501	NA	10
1967	11	594	13,112	NA	10
1968	10	609	14,211	NA	10
1969	7	651	14,814	NA	10
1970	7	745	15,311	NA	11
1971	5	766	15,923	NA	10
1972	4	787	16,892	NA	10
1973	3	743	17,829	NA	11
1974	2	685	17,400	NA	10
1975	1	595	17,615	NA	10
1976	(s)	559	18,506	NA	10
1977	(s)	543	19,240	NA	10
1978	(4)	539	20,040	NA	10
1979	(4)	612	19,823	NA	10
1980	(4)	650	19,007	NA	11
1981	(4)	658	18,810	7	11
1982	(4)	612	18,419	19	11
1983	(4)	505	18,591	35	13
1984	(4)	545	19,218	43	14
1985	(4)	519	19,505	52	14
1986	(4)	499	20,269	60	15
1987	(4)	535	20,870	69	16
1988	(4)	632	21,629	70	16
1989	(4)	649	21,867	71	16
1990	(4)	680	21,809	63	16
1991	(4)	620	21,456	73	16
1992	(4)	606	21,812	83	16
1993	(4)	643	22,199	97	16
1994	(4)	707	22,761	109	17
1995	(4)	722	23,199	117	17
1996	(4)	734	R23,734	84	17
1997	(4)	776	R23,992	106	17
1998	(4)	662	R24,677	117	17
1999	(4)	R762	R25,493	122	17
2000P	(4)	774	25,807	139	18

[1]Natural gas consumed in the operation of pipelines (primarily in compressors) and small amounts consumed as vehicle fuel. See Table 6.5.
[2]Alcohol (ethanol blended into motor gasoline) is included in both "Petroleum" and "Alcohol Fuels," but is counted only once in both total primary consumption and total consumption.
[3]Electric utility retail sales of electricity, including nonutility sales of electricity to utilities for distribution to end users; beginning in 1996, also includes sales to ultimate consumers by power marketers.
[4]Since 1978, the small amounts of coal consumed for transportation are reported as industrial sector consumption.
R=Revised. P=Preliminary. NA=Not available. (s)=Less than 0.5 trillion Btu.

SOURCE: "Table 2.1e Transportation Sector Energy Consumption, 1949–2000 (Trillion Btu)," in *Annual Energy Review 2000*, U.S. Department of Energy, Energy Information Administration, Washington, DC, 2001

TABLE 1.8

Electric power sector energy consumption, 1949–2000

(trillion btu)

	Primary Consumption									
	Fossil Fuels				Renewable Energy[1]					
Year	Coal	Natural Gas[2]	Petroleum	Nuclear Electric Power	Conventional Hydroelectric Power[3]	Wood	Waste	Geothermal[4]	Solar	Wind
1949	1,995	569	415	0	1,449	6	NA	NA	NA	NA
1950	2,199	651	472	0	1,440	5	NA	NA	NA	NA
1951	2,507	791	400	0	1,454	5	NA	NA	NA	NA
1952	2,557	942	420	0	1,496	6	NA	NA	NA	NA
1953	2,777	1,070	514	0	1,439	5	NA	NA	NA	NA
1954	2,841	1,206	417	0	1,388	3	NA	NA	NA	NA
1955	3,458	1,194	471	0	1,407	3	NA	NA	NA	NA
1956	3,790	1,283	455	0	1,487	2	NA	NA	NA	NA
1957	3,855	1,383	498	(s)	1,557	2	NA	NA	NA	NA
1958	3,721	1,421	486	2	1,629	2	NA	NA	NA	NA
1959	4,029	1,686	552	2	1,587	2	NA	NA	NA	NA
1960	4,228	1,785	553	6	1,657	2	NA	1	NA	NA
1961	4,355	1,889	557	20	1,680	1	NA	2	NA	NA
1962	4,622	2,035	560	26	1,822	1	NA	2	NA	NA
1963	5,050	2,211	585	38	1,772	1	NA	4	NA	NA
1964	5,380	2,397	634	40	1,907	2	NA	5	NA	NA
1965	5,821	2,395	722	43	2,058	3	NA	4	NA	NA
1966	6,302	2,696	883	64	2,073	3	NA	4	NA	NA
1967	6,445	2,834	1,011	88	2,344	3	NA	7	NA	NA
1968	6,994	3,245	1,181	142	2,342	4	NA	9	NA	NA
1969	7,219	3,596	1,571	154	2,659	3	NA	13	NA	NA
1970	7,227	4,054	2,117	239	2,654	1	2	11	NA	NA
1971	7,299	4,099	2,495	413	2,861	1	2	12	NA	NA
1972	7,811	4,084	3,097	584	2,944	1	2	31	NA	NA
1973	8,658	3,748	3,515	910	3,010	1	2	43	NA	NA
1974	8,534	3,519	3,365	1,272	3,309	1	2	53	NA	NA
1975	8,786	3,240	3,166	1,900	3,219	(s)	2	70	NA	NA
1976	9,720	3,152	3,477	2,111	3,066	1	2	78	NA	NA
1977	10,262	3,284	3,901	2,702	2,515	3	2	77	NA	NA
1978	10,238	3,297	3,987	3,024	3,141	2	1	64	NA	NA
1979	11,260	3,613	3,283	2,776	3,141	3	2	84	NA	NA
1980	12,123	3,810	2,634	2,739	3,118	3	2	110	NA	NA
1981	12,583	3,768	2,202	3,008	3,105	3	1	123	NA	NA
1982	12,582	3,342	1,568	3,131	3,572	2	1	105	NA	NA
1983	13,213	2,998	1,544	3,203	3,899	2	2	129	NA	(s)
1984	14,019	3,220	1,286	3,553	3,800	5	4	165	(s)	(s)
1985	14,542	3,160	1,090	4,149	3,398	8	7	198	(s)	(s)
1986	14,444	2,691	1,452	4,471	3,446	5	7	219	(s)	(s)
1987	15,173	2,935	1,257	4,906	3,117	8	7	229	(s)	(s)
1988	15,850	2,709	1,563	5,661	2,662	10	8	217	(s)	(s)
1989	15,988	2,871	1,685	5,677	R,13,014	1289	1104	R,1325	17	124
1990	16,190	2,882	1,250	6,162	R3,146	316	137	R344	7	32
1991	16,028	2,856	1,178	6,580	R3,159	346	164	R352	8	32
1992	16,211	2,826	951	6,608	R2,818	368	184	R362	8	30
1993	16,790	2,741	1,052	6,520	R3,119	379	191	R374	9	31
1994	16,895	3,053	968	6,838	R2,993	390	197	R378	8	36
1995	16,990	3,276	658	7,177	R3,481	375	209	R319	8	33
1996	17,953	2,798	725	7,168	R3,892	380	214	R331	9	35
1997	18,501	3,025	822	6,678	R3,961	355	213	R306	9	33
1998	18,685	3,330	1,166	7,157	R3,569	329	220	R310	9	31
1999	19,533	5,811	1,349	R7,736	R3,512	389	243	R354	9	46
2000P	20,503	6,475	1,209	8,009	3,107	409	254	298	9	51

[1]Beginning in 1989, includes expanded coverage of nonutility consumption.
[2]Includes supplemental gaseous fuels.
[3]Through 1988, includes all electricity net imports. From 1989, includes electricity net imports derived from hydroelectric power only.
[4]From 1989, includes electricity imports from Mexico that are derived from geothermal energy.
R=Revised. P=Preliminary. (s)=Less than 0.5 trillion Btu. NA=Not available.
Web Page: http://www.eia.doe.gov/fuelrenewable.html.

SOURCE: "Table 2.1f Electric Power Sector Energy Consumption, 1949–2000 (Trillion Btu)," in *Annual Energy Review 2000*, U.S. Department of Energy, Energy Information Administration, Washington, DC, 2001

TABLE 1.9

World primary energy production by source, 1970–1999

(quadrillion btu)

Year	Coal	Natural Gas[1]	Crude Oil[2]	Natural Gas Plant Liquids	Nuclear Electric Power [3]	Hydroelectric Power[3]	Geothermal[3] and Other[4]	Total
1970	62.96	37.09	97.09	3.61	0.90	12.15	1.59	215.39
1971	61.72	39.80	102.70	3.85	1.23	12.74	1.61	223.64
1972	63.65	42.08	108.52	4.09	1.66	13.31	1.68	234.99
1973	63.87	44.44	117.88	4.23	2.15	13.52	1.73	R247.83
1974	63.79	45.35	117.82	4.22	2.86	14.84	R1.76	250.64
1975	66.20	45.67	113.08	4.12	3.85	15.03	1.74	249.69
1976	67.32	47.62	122.92	4.24	4.52	15.08	1.97	263.67
1977	68.46	48.85	127.75	4.40	5.41	15.56	R2.11	272.54
1978	69.56	50.26	128.51	4.55	6.42	16.80	2.32	278.41
1979	73.83	53.93	133.87	4.87	6.69	17.69	2.48	293.36
1980	R72.72	54.73	128.12	5.10	7.58	18.06	2.95	R289.26
1981	R73.04	55.56	120.16	5.36	8.53	18.35	3.09	R284.09
1982	R75.64	55.49	114.51	5.34	9.51	18.83	3.24	R282.56
1983	R75.70	R56.12	113.97	5.34	10.72	19.73	3.51	R285.10
1984	R79.86	61.78	116.86	5.71	12.99	20.35	3.64	R301.19
1985	R83.64	64.22	115.40	5.82	15.37	20.57	3.67	R308.68
1986	R85.76	65.32	120.24	6.12	16.34	21.03	R3.73	R318.55
1987	R87.56	R68.48	121.16	6.32	17.80	21.10	R3.79	R326.21
1988	R89.22	R71.80	125.93	6.63	19.30	21.90	3.94	R338.73
1989	R90.61	74.24	127.98	6.67	19.82	21.76	R4.30	R345.38
1990	R91.87	75.91	129.50	6.85	20.37	22.57	R3.96	R351.03
1991	R87.11	76.68	128.77	7.13	21.29	R22.99	R4.04	R348.01
1992	R87.87	R76.89	129.13	7.38	21.36	R22.94	R4.33	R349.89
1993	R85.19	R78.40	128.86	7.67	22.07	R24.30	R4.37	R350.84
1994	R86.76	79.16	130.46	7.84	22.50	R24.48	R4.60	R355.80
1995	R89.24	80.23	133.32	8.14	R23.31	R25.71	R4.76	R364.72
1996	R89.24	R84.06	136.64	8.30	R24.13	R26.10	R4.93	R373.39
1997	R92.51	R84.01	140.52	8.49	R23.90	R26.73	R4.95	R381.13
1998	R89.58	R85.56	R143.15	R8.74	R24.41	R26.68	R5.10	R383.21
1999P	84.76	87.31	140.84	8.87	25.25	27.18	5.54	379.75

[1]Dry production.
[2]Includes lease condensate.
[3]Net generation, i.e., gross generation less plant use.
[4]Includes net electricity generation from wood, waste, solar, and wind. Data for the United States also include other renewable energy.
R=Revised. P=Preliminary.
Notes: Totals may not equal sum of components due to independent rounding.
World primary energy production includes production of crude oil (including lease condesate), natural gas plant liquids, dry natural gas, and coal; and net electricity generation from hydroelectric power, nuclear electric power, geothermal, wood, waste, solar, and wind. Data for the United States also include wood, waste, geothermal, and solar energy not used for electricity generation. Crude oil production is measured at the wellhead and includes lease condensate. Natural gas plant liquids are products obtained from processing natural gas at gas processing plants, including natural gas plants, cycling plants, and fractionators. Dry natural gas production is that amount of natural gas produced that is available to be marketed and consumed as a gas. Coal (anthracite, bituminous, subbituminous, and lignite) production is the sum of sales, mine consumption, issues to miners, and issues to coking, briquetting, and other ancillary plants at mines. Coal production data include quantities extracted from surface and under ground mines and normally exclude wastes removed at mines or associated preparation plants. The data on generation of electricity from hydroelectric power, nuclear electric power, wood, waste, geothermal, solar, and wind include data on both electric utility and nonutility generation reported on a net basis, thus excluding electricity that is generally used by the electric power plant for its own operating purposes or elecricity losses in the transformers that are considered integral parts of the station.
Web Page: http://www.eia.doe.gov/international.

SOURCE: "Table 11.2 World Primary Energy Production by Source, 1970–1999," in *Annual Energy Review 2000*, U.S. Department of Energy, Energy Information Administration, Washington, DC, 2001

such as solar, wind, waste, and geothermal energies are used very little to generate the nation's electricity.

INTERNATIONAL ENERGY USAGE

World Production

World production of primary energy rose from 248 quadrillion Btu in 1973 to 380 quadrillion Btu in 1999. (See Table 1.9.) The Energy Information Administration (EIA) of the U.S. Department of Energy stated in its 2001 report *Annual Energy Review 2000* that the world's total output of primary energy increased by 76 percent from 1970 to 1999. In 1999 fossil fuels were the most heavily produced fuel, accounting for 85 percent of all energy produced worldwide. Renewable energy accounted for 9 percent of all energy produced worldwide that year, and nuclear power accounted for 7 percent.

In 1999 the United States, Russia, and China were, by far, the leading producers of energy, followed by Saudi Arabia, Canada, and the United Kingdom. (See Figure 1.11.) Almost all the energy from the Middle East is in the form of oil or natural gas, while coal is a major source in China. Canada is the leading producer of hydroelectric power and alone accounts for 14 percent of world production. France produces the highest percentage of its energy from nuclear power.

FIGURE 1.11

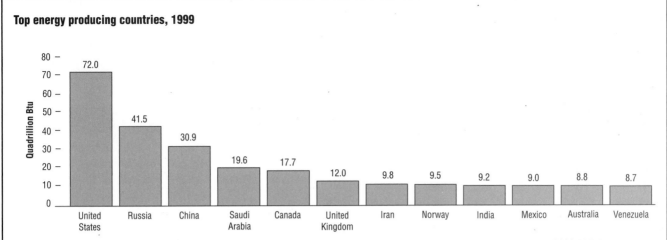

Top energy producing countries, 1999

SOURCE: Adapted from "Figure 11.1 World Primary Energy Production: Top Producing Countries, 1999," in *Annual Energy Review 2000*, U.S. Department of Energy, Energy Information Administration, Washington, DC, 2001

World Consumption

Table 1.10 shows the world consumption of energy by country and region from 1991 to 2000. Five countries— the United States, China, Russia, Japan, and Germany— together consumed 50 percent of the world's total energy in 2000. The United States, by far the world's largest consumer of energy, used 98.8 quadrillion Btu, or nearly one-fourth of energy consumed worldwide. This amount was about 2.7 times higher than China's 36.7 quadrillion Btu, while Russia consumed 28.1 quadrillion Btu.

FUTURE TRENDS IN ENERGY CONSUMPTION, PRODUCTION, AND PRICES

The Energy Information Administration (EIA) of the U.S. Department of Energy forecasts energy supply, demand, and prices every year in its *Annual Energy Outlook,* which is used by decision makers in the public and private sectors. The EIA's latest projections, from 2000 through 2020, are based on current U.S. laws, regulations, and economic conditions.

Total energy consumption is projected to increase from 99.3 to 130.9 quadrillion Btu between 2000 and 2020, an average annual increase of 1.4 percent. (See Table 1.11.) That projection becomes higher with high economic growth and/or low world oil prices, and lower with low economic growth and/or high world oil prices.

Energy consumption will increase in all end-use sectors to 2020. (See Figure 1.12.) According to *Annual Energy Outlook 2002,* consumption in the transportation sector is projected to increase the most—1.9 percent per year from 2000 to 2020. In 1998 light vehicles, which include cars, light trucks, sport utility vehicles (SUVs), and vans, accounted for 63 percent of energy use in the transportation sector. (See Figure 1.13.) The energy demand for

FIGURE 1.12

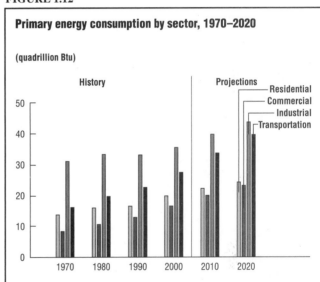

Primary energy consumption by sector, 1970–2020

(quadrillion Btu)

SOURCE: "Figure 26. Primary energy consumption by sector, 1970–2020 (quadrillion Btu)," in *Annual Energy Outlook 2002,* U.S. Department of Energy, Energy Information Administration, Washington, DC, 2001

these vehicles is projected to rise because of rapid growth in travel coupled with slower growth in fuel efficiency.

Following the transportation sector, energy consumption in the commercial sector is projected to grow next fastest, at an average annual rate of 1.7 percent. (See Figure 1.12.) This projected increase is expected from increased computer, office equipment, and telecommunications use. Residential and industrial energy demands are each expected to grow less—about 1 percent annually. Contributing to the projected slow growth of energy consumption in industry is the continuing shift to less energy-intensive manufacturing. The projected growth in residential energy consumption is due to increased home use of computers, electronic equipment, and appliances.

TABLE 1.10

World primary energy consumption (Btu), 1991–2000

Region[1] Country	1991	1992	1993	1994	1995	1996	1997	1998	1999	2000
North America										
Canada	10.89	10.94	11.46	11.74	11.75	12.11	12.37	12.20	12.74	13.07
Mexico	5.02	5.12	5.13	5.30	5.31	5.55	5.65	5.93	6.06	6.18
United States	84.33	85.55	87.33	89.25	90.98	93.97	94.38	94.66	96.77	98.79
Other	0.01	0.02	0.02	0.02	0.02	0.02	0.02	0.02	0.02	0.02
Total	**100.26**	**101.63**	**103.94**	**106.31**	**108.05**	**111.64**	**112.41**	**112.80**	**115.58**	**118.05**
Central & South America										
Argentina	1.99	2.12	2.29	2.32	2.40	2.47	2.57	2.73	2.59	2.71
Brazil	6.21	6.30	6.58	6.89	7.30	7.76	8.19	8.45	8.72	9.10
Chile	0.56	0.60	0.65	0.69	0.75	0.82	0.95	0.94	0.99	1.03
Colombia	0.97	0.98	1.07	1.10	1.12	1.18	1.23	1.25	1.18	1.18
Cuba	0.46	0.41	0.40	0.41	0.42	0.43	0.39	0.37	0.38	0.39
Venezuela	2.21	2.22	2.29	2.42	2.47	2.58	2.66	2.81	2.72	2.72
Other	2.64	2.72	2.84	3.03	3.23	3.30	3.47	3.71	3.80	3.93
Total	**15.04**	**15.35**	**16.12**	**16.87**	**17.69**	**18.54**	**19.46**	**20.27**	**20.39**	**21.07**
Western Europe										
Austria	1.23	1.19	1.23	1.23	1.28	1.29	1.33	1.35	1.44	1.41
Belgium	2.27	2.24	2.26	2.31	2.36	2.55	2.63	2.66	2.61	2.75
Denmark	0.83	0.82	0.85	0.84	0.88	0.88	0.91	0.90	0.88	0.88
Finland	1.14	1.18	1.20	1.23	1.12	1.14	1.26	1.29	1.30	1.30
France	9.39	9.41	9.37	9.28	9.54	9.92	9.87	10.19	10.30	10.41
Germany	14.31	14.00	14.06	14.01	14.32	14.30	14.30	14.33	14.13	13.98
Greece	1.07	1.04	1.09	1.11	1.12	1.15	1.22	1.29	1.29	1.33
Ireland	0.39	0.40	0.40	0.42	0.45	0.47	0.49	0.53	0.56	0.59
Italy	7.17	7.22	7.05	6.97	7.56	7.64	7.45	7.73	7.77	7.96
Netherlands	3.56	3.53	3.60	3.57	3.70	3.82	3.83	3.81	3.83	3.91
Norway	1.59	1.65	1.65	1.66	1.73	1.74	1.81	1.86	1.89	1.79
Portugal	0.75	0.76	0.78	0.81	0.85	0.88	0.94	0.99	1.01	1.08
Spain	4.15	4.12	4.04	4.22	4.48	4.39	4.72	5.02	5.21	5.40
Sweden	2.17	2.17	2.18	2.19	2.34	2.28	2.18	2.28	2.23	2.25
Switzerland	1.21	1.21	1.20	1.20	1.17	1.21	1.23	1.21	1.23	1.24
Turkey	2.08	2.13	2.33	2.23	2.47	2.74	2.96	3.02	2.92	3.20
United Kingdom	9.60	9.33	9.65	9.64	9.60	10.16	9.88	9.87	9.79	9.88
Former Yugoslavia	1.87	—	—	—	—	—	—	—	—	—
Croatia	—	0.33	0.33	0.36	0.37	0.37	0.38	0.40	0.39	0.41
Yugoslavia	—	0.69	0.55	0.60	0.46	0.70	0.73	0.77	0.64	0.59
Other	0.31	0.81	0.81	0.79	0.87	0.85	0.84	0.89	0.91	0.93
Total	**65.08**	**64.23**	**64.63**	**64.69**	**66.67**	**68.49**	**68.96**	**70.37**	**70.32**	**71.29**
Eastern Europe & Former U.S.S.R.										
Bulgaria	1.01	1.00	0.93	0.92	0.99	1.01	0.96	0.90	0.83	0.94
Former Czechoslovakia	3.55	3.24	—	—	—	—	—	—	—	—
Czech Republic	—	—	1.66	1.56	1.65	1.80	1.72	1.56	1.48	1.45
Slovakia	—	—	0.79	0.77	0.82	0.81	0.80	0.79	0.80	0.78
Hungary	1.16	1.08	1.06	1.06	1.06	1.09	1.07	1.07	1.07	1.05
Poland	3.88	3.87	4.00	3.84	3.69	3.55	4.09	3.83	3.68	3.68
Romania	2.24	2.06	1.99	1.88	2.02	2.06	2.03	1.75	1.56	1.5
Former U.S.S.R	57.46	—	—	—	—	—	—	—	—	—
Azerbaijan	—	0.99	0.85	0.76	0.73	0.65	0.64	0.55	0.57	0.53
Belarus	—	1.57	1.34	1.10	1.06	1.07	1.07	1.07	1.06	1.08
Kazakhstan	—	3.37	2.79	2.25	2.04	1.99	1.70	1.64	1.53	1.79
Lithuania	—	0.44	0.37	0.36	0.37	0.32	0.33	0.35	0.28	0.27
Russia	—	34.88	32.67	29.63	28.24	27.92	25.52	25.62	27.45	28.07
Turkmenistan	—	0.29	0.27	0.27	0.29	0.28	0.29	0.25	0.31	0.37
Ukraine	—	8.89	8.58	7.31	7.21	6.73	6.44	6.26	6.41	6.46
Uzbekistan	—	1.66	2.04	1.76	1.85	1.91	1.89	1.84	1.87	1.92
Other	0.09	1.83	1.39	1.20	1.17	1.28	1.23	1.27	1.18	1.17
Total	**69.40**	**65.16**	**60.75**	**54.66**	**53.20**	**52.47**	**49.78**	**48.76**	**50.08**	**51.14**

The consumption of petroleum, natural gas, coal, and non-hydroelectric renewable energy sources is expected to rise significantly from 2000 to 2020. (See Figure 1.14.) Worldwide oil consumption is projected to increase from 76 million barrels of oil per day in 2000 to 118.9 million barrels per day in 2020. Coal, natural gas, and renewable fuels consumption is projected to grow in part to meet the increased demand for electricity. Hydroelectric power

consumption will remain steady, since no new dams are projected. Electricity generated from nuclear power will decline slightly through 2020, with no new plants being built and old plants being retired.

The rising consumption of petroleum is projected to lead to increasing petroleum imports by the United States through 2020. (See Figure 1.6.) OPEC oil production is expected to nearly double, from 30.9 million barrels per

TABLE 1.10

World primary energy consumption (Btu), 1991–2000 [CONTINUED]

Region[1] Country	1991	1992	1993	1994	1995	1996	1997	1998	1999	2000
Middle East										
Bahrain	0.29	0.24	0.29	0.28	0.29	0.29	0.35	0.36	0.36	0.37
Iran	3.23	3.35	3.47	3.66	3.81	3.95	4.44	4.47	4.61	4.72
Iraq	0.60	0.84	0.96	1.08	1.13	1.12	1.03	1.05	1.07	1.09
Israel	0.48	0.54	0.60	0.62	0.61	0.65	0.70	0.75	0.76	0.78
Kuwait	0.21	0.35	0.48	0.57	0.59	0.75	0.80	0.85	0.92	0.99
Oman	0.21	0.20	0.23	0.24	0.22	0.23	0.27	0.35	0.30	0.34
Qatar	0.41	0.49	0.57	0.57	0.58	0.59	0.64	0.65	0.62	0.66
Saudi Arabia	3.28	3.39	3.52	3.64	3.85	4.05	4.08	4.27	4.35	4.57
Syria	0.56	0.59	0.64	0.68	0.65	0.70	0.74	0.81	0.83	0.82
United Arab Emirates	1.49	1.55	1.48	1.49	1.60	1.67	1.79	1.84	1.86	1.74
Yemen	0.17	0.17	0.14	0.14	0.14	0.14	0.15	0.14	0.13	0.14
Other	0.33	0.36	0.39	0.45	0.47	0.49	0.52	0.54	0.56	0.57
Total	**11.26**	**12.06**	**12.77**	**13.41**	**13.97**	**14.65**	**15.50**	**16.08**	**16.37**	**16.80**
Africa										
Algeria	1.35	1.29	1.20	1.23	1.30	1.26	1.20	1.25	1.25	1.23
Angola	0.09	0.09	0.09	0.09	0.09	0.09	0.10	0.09	0.10	0.09
Egypt	1.43	1.43	1.51	1.55	1.58	1.73	1.80	1.87	1.91	2.04
Gabon	0.04	0.05	0.04	0.05	0.05	0.06	0.06	0.05	0.05	0.05
Libya	0.53	0.49	0.51	0.53	0.54	0.57	0.59	0.57	0.53	0.58
Morocco	0.32	0.33	0.36	0.40	0.37	0.39	0.40	0.41	0.44	0.42
Nigeria	0.77	0.78	0.80	0.74	0.83	0.85	0.85	0.81	0.81	0.83
South Africa	3.58	3.73	3.80	3.85	4.16	3.97	4.64	4.88	4.60	4.64
Zimbabwe	0.23	0.24	0.22	0.23	0.23	0.23	0.22	0.21	0.24	0.25
Other	1.41	1.48	1.51	1.58	1.57	1.64	1.69	1.70	1.75	1.75
Total	**9.76**	**9.91**	**10.05**	**10.25**	**10.73**	**10.78**	**11.57**	**11.84**	**11.68**	**11.88**
Asia & Oceania										
Australia	3.70	3.82	3.93	3.96	4.11	4.18	4.56	4.60	4.84	4.89
Bangladesh	0.26	0.29	0.31	0.34	0.37	0.39	0.40	0.42	0.47	0.50
Brunei	0.04	0.05	0.05	0.05	0.06	0.06	0.07	0.06	0.07	0.08
China	28.26	29.31	31.36	34.04	35.21	36.04	37.61	37.07	37.02	36.67
Hong Kong	0.47	0.52	0.56	0.61	0.66	0.68	0.52	0.68	0.89	0.80
India	8.06	8.71	9.10	9.59	11.10	11.25	11.55	11.78	12.12	12.67
Indonesia	2.36	2.54	2.87	3.06	3.26	3.52	3.68	3.52	3.72	3.85
Japan	18.89	19.14	19.41	20.18	20.83	21.48	21.78	21.43	21.57	21.77
Korea, North	3.05	3.02	3.12	3.08	3.04	2.97	2.82	2.72	2.76	2.81
Korea, South	4.28	4.79	5.55	6.01	6.62	6.95	7.40	6.82	7.31	7.88
Malaysia	1.09	1.14	1.29	1.43	1.47	1.64	1.66	1.68	1.74	1.86
New Zealand	0.73	0.74	0.77	0.80	0.86	0.82	0.80	0.79	0.80	0.83
Pakistan	1.25	1.29	1.41	1.50	1.58	1.70	1.68	1.73	1.81	1.91
Philippines	0.73	0.77	0.84	0.90	0.96	1.02	1.09	1.13	1.18	1.23
Singapore	0.86	0.97	1.08	1.16	1.18	1.35	1.49	1.54	1.62	1.68
Taiwan	2.09	2.21	2.43	2.63	2.96	3.16	3.27	3.42	3.69	3.78
Thailand	1.37	1.47	1.68	1.87	2.25	2.44	2.52	2.37	2.50	2.56
Vietnam	0.28	0.30	0.38	0.41	0.51	0.55	0.54	0.55	0.64	0.68
Other	0.57	0.55	0.58	0.63	0.64	0.65	0.66	0.69	0.72	0.74
Total	**78.33**	**81.61**	**86.73**	**92.24**	**97.67**	**100.84**	**104.09**	**102.99**	**105.48**	**107.16**
World Total	**349.14**	**349.96**	**354.98**	**358.43**	**367.99**	**377.42**	**381.77**	**383.12**	**389.89**	**397.40**

[1] Preliminary.
-- = Not applicable.
(s) = Value less than 5 trillion Btu.
Notes: Sum of components may not equal total due to independent rounding. Primary energy consumption reported in this table includes petroleum, dry natural gas, coal, net hydroelectric, nuclear, geothermal, solar, wind, and wood and waste electric power. Primary energy consumption for the United States also includes: (1) the consumption of geothermal, solar, and wood and waste energy not used for electricity generation; (2) electricity imports from Mexico that are derived from geothermal energy; and (3) net imports of electricity derived from nonrenewable sources. Primary energy consumption for all countries, except the United States, has been adjusted to include total electricity imports and to exclude total electricity exports. This adjustment is necessary because the consumption data for electric power by type, are not adjusted for electricity imports and exports, except for hydroelectric power in the United States.
As a result of these adjustments, primary energy consumption reported in this table might not be equal to sum of the individual fuel types reported.

SOURCE: "Table E1. World Primary Energy Consumption (Btu), 1991–2000," in *International Energy Annual 2000,* U.S. Department of Energy, Energy Information Administration, Washington, DC, 2002

day in 2000 to 57.5 million barrels per day in 2020. Non-OPEC oil production is expected to increase from 45.7 million barrels per day in 2000 to 61.1 million barrels per day in 2020. Global oil production is estimated to peak between 2011 and 2025. World oil production per person has been declining since 1979, as populations are growing faster than oil production, and this trend is expected to continue.

In the period 2000–20, electricity prices in the United States are projected to decline slightly because of restructuring laws designed to increase competition in the industry,

TABLE 1.11

Summary of results for five scenarios

			2020				
Sensitivity factors	1999	2000	Reference	Low Economic Growth	High Economic Growth	Low World Oil Price	High World Oil Price
Primary production (quadrillion Btu)							
Petroleum	15.06	15.04	15.95	15.52	16.39	14.40	17.73
Natural gas	19.20	19.59	29.25	27.98	29.72	28.54	30.03
Coal	23.15	22.58	28.11	26.88	30.08	27.58	29.04
Nuclear power	7.74	8.03	7.49	7.38	7.49	7.31	7.58
Renewable energy	6.69	6.46	8.93	8.59	9.37	8.90	8.97
Other	1.66	1.10	0.93	0.91	0.73	0.40	1.06
Total primary production	**73.50**	**72.80**	**90.66**	**87.26**	**93.79**	**87.13**	**94.40**
Net imports (quadrillion Btu)							
Petroleum (including SPR)	21.19	22.28	35.04	32.39	38.25	38.65	31.51
Natural gas	3.50	3.60	5.64	5.12	6.40	5.90	5.17
Coal/Other (- indicates export)	-0.96	-0.77	-0.29	-0.38	-0.15	-0.31	-0.29
Total net imports	**23.73**	**25.11**	**40.39**	**37.13**	**44.49**	**44.25**	**36.40**
Discrepancy	0.13	-1.37	0.20	0.25	0.04	-0.04	0.51
Consumption (quadrillion Btu)							
Petroleum products	38.25	38.63	51.99	48.84	55.60	53.78	50.96
Natural gas	22.57	23.43	34.63	32.84	35.87	34.17	34.04
Coal	21.56	22.34	27.35	26.08	29.41	26.83	28.27
Nuclear power	7.74	8.03	7.49	7.38	7.49	7.31	7.58
Renewable energy	6.70	6.48	8.94	8.59	9.38	8.91	8.98
Other	0.28	0.38	0.44	0.40	0.48	0.42	0.46
Total consumption	**97.10**	**99.29**	**130.85**	**124.13**	**138.24**	**131.42**	**130.29**
Prices (2000 dollars)							
World oil price (dollars per barrel)	17.60	27.72	24.68	23.45	25.81	17.64	30.58
Domestic natural gas at wellhead (dollars per thousand cubic feet)	2.27	3.60	3.26	2.94	3.65	3.07	3.40
Domestic coal at minemouth (dollars per short ton)	17.01	16.45	12.79	12.56	13.23	12.67	12.95
Average electricity price (cents per kilowatthour)	6.7	6.9	6.5	6.2	6.8	6.4	6.5
Economic indicators							
Real Gross Domestic Product (billion 1996 dollars)	8,857	9,224	16,525	14,901	18,102	16,561	16,496
(annual change, 2000-2020)	—	—	3.0%	2.4%	3.4%	3.0%	2.9%
GDP Chain-Type Price Index (index, 1996=1.00)	1.047	1.070	1.826	2.067	1.608	1.797	1.859
(annual change, 2000-2020)	—	—	2.7%	3.3%	2.1%	2.6%	2.8%
Real disposable personal income (billion 1996 dollars)	6,320	6,539	11,698	10,791	12,541	11,685	11,723
(annual change, 2000-2020)	—	—	3.0%	2.5%	3.3%	2.9%	3.0%
Gross manufacturing output (billion 1992 dollars)	3,804	4,022	7,003	6,473	8,023	7,026	6,977
(annual change, 2000-2020)	—	—	2.8%	2.4%	3.5%	2.8%	2.8%
Energy intensity							
(thousand Btu per 1996 dollar of GDP)	10.97	10.77	7.92	8.34	7.64	7.94	7.90
(annual change, 2000-2020)	—	—	-1.5%	-1.3%	-1.7%	-1.5%	-1.5%
Carbon dioxide emissions							
(million metric tons carbon equivalent)	1,517	1,562	2,088	1,980	2,215	2,103	2,083
(annual change, 2000-2020)	—	—	1.5%	1.2%	1.8%	1.5%	1.5%

Notes: Quantities are derived from historical volumes and assumed thermal conversion factors. Other production includes liquid hydrogen, methanol, supplemental natural gas, and some inputs to refineries. Net imports of petroleum include crude oil, petroleum products, unfinished oils, alcohols, ethers, and blending components. Other net imports include coal coke and electricity. Some refinery inputs appear as petroleum product consumption. Other consumption includes net electricity imports, liquid hydrogen, and methanol.

SOURCE: "Table 1. Summary of results for five cases," in *Annual Energy Outlook 2002,* U.S. Department of Energy, Energy Information Administration, Washington, DC, 2001. EIA data sources include DRI-WEFA and various U.S. government publications.

though restructuring has slowed due to problems with it in California. As shown in Figure 1.15, the electricity price projections in *Annual Energy Outlook 2002* (AEO2002) are higher than the projections from *Annual Energy Outlook 2001* (AEO2001) because they include the contracts entered into by California to guarantee electricity supplies in the state. Figure 1.15 also shows that coal prices are expected to decline from 2000 to 2020, while crude oil and natural gas prices are expected to decrease from 2000 levels, then slowly rise through 2020.

FIGURE 1.13

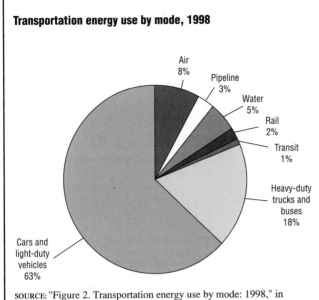

Transportation energy use by mode, 1998

SOURCE: "Figure 2. Transportation energy use by mode: 1998," in *Transportation Statistics Annual Report 2000*, U.S. Department of Transportation, Bureau of Transportation Statistics, Washington, DC, 2001. Compiled from various sources, as cited in U.S. Department of Transportation's *National Transportation Statistics 2000*.

Conflicting Priorities?: Energy and the Environment

Historically, the production, consumption, and price of energy have been driven largely by the supply of and demand for petroleum, but environmental concerns may play a greater role in the future.

Whether the American public will adjust its energy use out of concern for the environment remains unclear. In 2001 the United States rejected the Kyoto Protocol to the United Nations Framework Convention on Climate Change (UNFCCC), an international treaty designed to curb global warming and air pollution. As an alternative, President Bush called for voluntary measures to slow the growth of emissions in the United States. The Bush administration holds that the timetables of the Kyoto treaty would harm the economy and that all nations must address global warming. As written in 2002, the Kyoto treaty required no action by large developing countries like China and India, whose emissions could soon surpass those of developed nations.

FIGURE 1.14

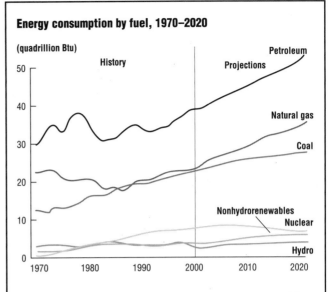

Energy consumption by fuel, 1970–2020

SOURCE: "Figure 2. Energy consumption by fuel, 1970–2020 (quadrillion Btu)," in *Annual Energy Outlook 2002*, U.S. Department of Energy, Energy Information Administration, Washington, DC, 2001

FIGURE 1.15

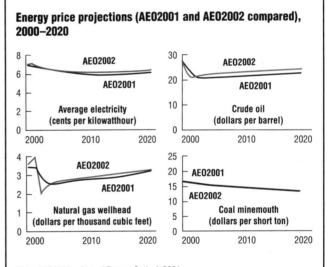

Energy price projections (AEO2001 and AEO2002 compared), 2000–2020

Note: AEO2001 = *Annual Energy Outlook 2001*
AEO2002 = *Annual Energy Outlook 2002*

SOURCE: "Figure 1. energy price projections, 2000–2020: AEO2001 and AEO2002 Compared (2000 dollars)," in *Annual Energy Outlook 2002*, U.S. Department of Energy, Energy Information Administration, Washington, DC, 2001

CHAPTER 2

OIL

THE QUEST FOR OIL

On August 27, 1859, Edwin Drake struck oil 69 feet below the surface of the earth near Titusville, Pennsylvania. This was the first successful modern oil well, which ushered in the "Age of Petroleum." Not only did petroleum help meet the growing demand for new and better fuels for heating and lighting, but it was also an excellent fuel for the internal combustion engine, which was developed in the late 1800s.

Sources of Oil

Almost all oil comes from underground reservoirs. The most widely accepted explanation of how oil and gas are formed within the earth is that these fuels are the products of intense heat and pressure applied over millions of years to organic (formerly alive) sediments buried in geological formations. For this reason they are called fossil fuels. They are limited resources—there is only so much fossil fuel.

At one time, it was believed that crude oil flowed in underground streams and accumulated in lakes or caverns in the earth. Today, scientists know that a petroleum reservoir is usually a solid sandstone or limestone formation overlaid with a layer of impermeable rock or shale, which creates a shield. The petroleum accumulates within the pores and fractures of the rock and is trapped beneath the seal. Anticlines (archlike folds in a bed of rock), faults, and salt domes are common trapping formations. (See Figure 2.1.) Oil and natural gas deposits can be found at varying depths. Wells are drilled to reach the reservoirs and extract the oil. Deep wells are more expensive to drill and are usually attempted only to reach large reservoirs or when the price of oil is high.

How Oil Is Collected

Oil is crucial to the economies of industrialized nations, and oil production is strongly influenced by mar-

FIGURE 2.1

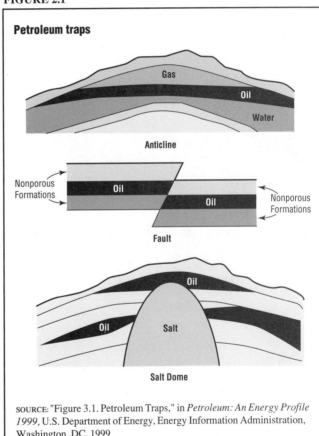

Petroleum traps

SOURCE: "Figure 3.1. Petroleum Traps," in *Petroleum: An Energy Profile 1999*, U.S. Department of Energy, Energy Information Administration, Washington, DC, 1999

ket prices. Scouting for new deposits and drilling exploratory wells is expensive. It is also costly to drill, maintain, and operate production wells. If the price of oil falls too low, operators will shut off expensive wells because they cannot recover their higher operating costs. In general, when oil prices are high, oil companies drill; when prices are low, drilling drops off.

Figure 2.2 shows a diagram of a rotary drilling system, or rotary rig. The rotating bit at the end of lengths of

FIGURE 2.2

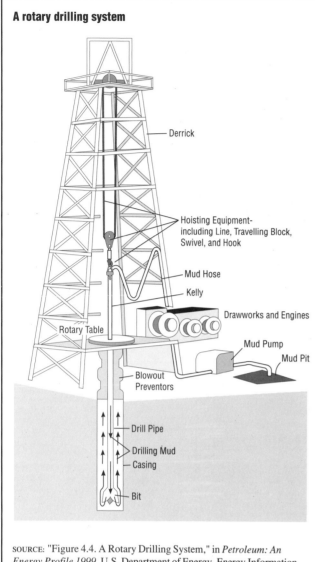

A rotary drilling system

Derrick

Hoisting Equipment-
including Line, Travelling Block,
Swivel, and Hook

Mud Hose

Kelly

Drawworks and Engines

Mud Pump

Mud Pit

Rotary Table

Blowout
Preventors

Drill Pipe

Drilling Mud

Casing

Bit

SOURCE: "Figure 4.4. A Rotary Drilling System," in *Petroleum: An Energy Profile 1999,* U.S. Department of Energy, Energy Information Administration, Washington, DC, 1999

pipe drills a hole into the ground. Drilling mud is pushed through the pipe and the drill bit, which forces the mud and bits of rock from the drilling process back to the surface, as shown by the "up" arrows in the diagram. As the well gets deeper, more pipe is added. The oil derrick supports equipment that can lift the pipe and drill bit from the well when drill bits need to be changed or replaced.

Oil deposits are not distributed evenly over the world, and some that have been exploited for decades are being exhausted. As an oil reservoir is depleted, various techniques can be used to recover additional petroleum. These include the injection of water, chemicals, or steam to force more oil from the rock. These recovery techniques can be expensive and are used only when the price of oil is relatively high.

In 1996, after a decade of low oil prices, the demand for rigs collapsed, and new rig construction stopped alto-

gether. Thousands of rigs were left idle, sold for scrap metal, or shipped overseas, and their crews were put out of work. Idle rigs became a source of spare parts for those still operating. In 1997, following a rise in oil prices, the demand for rigs soared, but by 1998 the market for rigs had once again dwindled as oil prices sank. Rising oil prices resulting from OPEC restrictions in 1999 eventually boosted the demand for drilling equipment. (OPEC is the Organization of Petroleum Exporting Countries. Current members are Algeria, Indonesia, Iran, Iraq, Kuwait, Libya, Nigeria, Qatar, Saudi Arabia, United Arab Emirates, and Venezuela.) Crude oil prices remained high as of October 2002.

TYPES OF OIL

While crude oil is usually dark when it comes from the ground, it may also be green, red, yellow, or colorless, depending on its chemical composition and the amount of sulfur, oxygen, nitrogen, and trace minerals present. Its viscosity (thickness, or resistance to flow) can range from as thin as water to as thick as tar. Crude oil is refined, or chemically processed, into finished petroleum products; it has limited uses in its natural form.

Crude oils vary in quality. "Sweet" crudes have little sulfur, refine easily, and are worth more than "sour" crudes, which contain more impurities. "Light" crudes have more short molecules, which yield more gasoline and are more profitable than "heavy" crudes, which have more long molecules and bring a lower price in the market.

In addition to crude oil, there are two other sources of petroleum: lease condensate and natural gas plant liquids. Lease condensate is a natural gas liquid recovered from gas well gas. It consists primarily of chemical compounds called penthanes and heavier hydrocarbons and is generally blended with crude oil for refining. Natural gas plant liquids are natural gas liquids that are recovered during the refinement of natural gas in processing plants.

HOW OIL IS REFINED

Before oil can be used by consumers, crude oil, lease condensate, and natural gas plant liquids must be processed into finished products. The first step in refining is distillation, in which crude oil molecules are separated according to size and weight.

During distillation, crude oil is heated until it turns to vapor. (See Figure 2.3.) The vaporized crude enters the bottom of a distillation column, where it rises and condenses on trays. The lightest vapors, such as those of gasoline, rise to the top. The middleweight vapors, such as those of kerosene, rise about halfway up the column. The heaviest vapors, such as those of heavy gas oil, stay at the bottom. The vapors at each level are then con-

FIGURE 2.3

Crude oil distillation

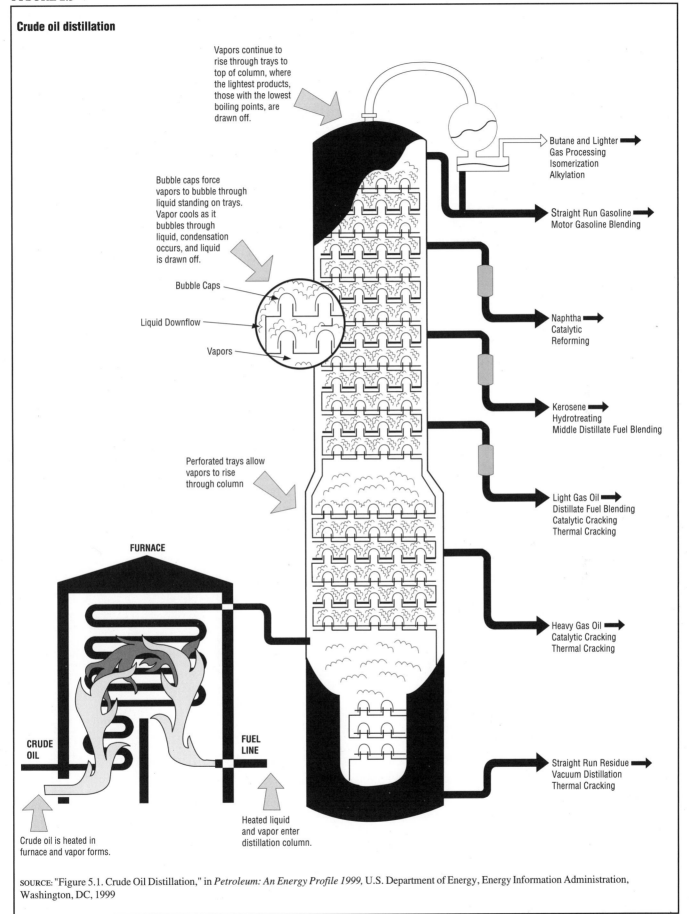

Vapors continue to rise through trays to top of column, where the lightest products, those with the lowest boiling points, are drawn off.

Bubble caps force vapors to bubble through liquid standing on trays. Vapor cools as it bubbles through liquid, condensation occurs, and liquid is drawn off.

Bubble Caps

Liquid Downflow

Vapors

Perforated trays allow vapors to rise through column

FURNACE

CRUDE OIL

FUEL LINE

Heated liquid and vapor enter distillation column.

Crude oil is heated in furnace and vapor forms.

Butane and Lighter
Gas Processing
Isomerization
Alkylation

Straight Run Gasoline
Motor Gasoline Blending

Naphtha
Catalytic
Reforming

Kerosene
Hydrotreating
Middle Distillate Fuel Blending

Light Gas Oil
Distillate Fuel Blending
Catalytic Cracking
Thermal Cracking

Heavy Gas Oil
Catalytic Cracking
Thermal Cracking

Straight Run Residue
Vacuum Distillation
Thermal Cracking

SOURCE: "Figure 5.1. Crude Oil Distillation," in *Petroleum: An Energy Profile 1999,* U.S. Department of Energy, Energy Information Administration, Washington, DC, 1999

densed to liquid as they are cooled. These liquids are drawn off, and the processes of cracking and reforming further refine each portion. Cracking converts the heaviest fractions of separated petroleum into lighter fractions to produce jet fuel, motor gasoline, home heating oil, and less-residual fuel oils, which are heavier and used for naval ships, commercial and industrial heating, and some power generation. Reforming is used to increase the octane rating of gasoline.

Refining is a continuous process, with crude oil entering the refinery at the same time as finished products leave by pipeline, truck, and train. Although storage tanks surround refineries, they have limited storage capacity. If there is a malfunction and products cannot be processed, they may be burned off (flared) if no storage facility is available. While a small flare is normal at a refinery or a chemical plant, a large flare, or many flares, likely indicates a processing problem.

HOW MANY REFINERIES ARE THERE?

In 2000, 158 refineries were operating in the United States, a drop from 336 in 1949 and 324 in 1981. (See Table 2.1.) Refinery capacity in 2000 was about 16.5 million barrels per day, below the 1981 peak of 18.6 million barrels. As of 2002 U.S. refineries were operating at near full capacity. Utilization rates generally increased from a low of 68.6 percent in 1981—a period of low demand because of economic recession—to a high of 95.6 percent in 1998. Although capacity fell slightly in 1999 and 2000 (to 92.6 percent each year), it is still high.

As Table 2.1 shows, fewer refineries operate in the United States today than in the past, but they are working at near maximum levels. One reason for the drop in the number of U.S. refineries is that the petroleum industry is shutting down older, inefficient refineries and concentrating production in more efficient plants, which are usually newer and larger. Consolidation within the industry has also played a role in refinery operation. For example, the merger of Gulf Oil Corporation into Chevron Corporation in 1984 led to the closing of two large refineries, one in Bakersfield, California, and the other in Cincinnati, Ohio. In 1998 Exxon merged with Mobil Oil and BP merged with Amoco. BP Amoco then bought out Arco in April 2000 to create the world's largest non-OPEC oil producer and the third-largest natural gas producer. Large merger possibilities continue in the industry.

Another reason for the drop in the number of U.S. refineries is that some OPEC countries have begun to refine their own oil products to maximize their profits. This strategy, employed particularly by Saudi Arabia, led to a drop in demand for U.S. refineries. As of November 2002 no major refinery manufacturer had plans to begin

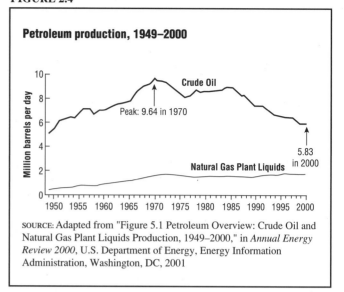

FIGURE 2.4

Petroleum production, 1949–2000

SOURCE: Adapted from "Figure 5.1 Petroleum Overview: Crude Oil and Natural Gas Plant Liquids Production, 1949–2000," in *Annual Energy Review 2000*, U.S. Department of Energy, Energy Information Administration, Washington, DC, 2001

construction, a process that takes three to five years to complete. The last large refinery to be built in the United States was completed in 1976, and the last refinery of any size to be completed began operation in Valdez, Alaska, in 1993.

USES FOR OIL

Many of the uses for petroleum are well known: gasoline, diesel fuel, jet fuel, and lubricants for transportation; heating oil, residual oil, and kerosene for heat; and heavy residuals for paving and roofing. Petroleum by-products are also vital to the chemical industry, ending up in many different foams, plastics, synthetic fabrics, paints, dyes, inks, and even pharmaceutical drugs. Many chemical plants, because of their dependence on petroleum, are directly connected by pipelines to nearby refineries.

The demand for petroleum products varies, and petroleum prices usually fluctuate with demand. Heating oil demand rises during the winter. A cold spell, which leads to a sharp rise in demand, may result in a corresponding price increase. A warm winter may be reflected in lower prices as suppliers try to clear out their inventory. Gasoline demand rises during the summer, when people drive more on vacations and for recreation, and gas prices usually rise as a consequence. Petroleum demand also reflects the general condition of the economy. During a recession, demand for and production of petroleum products drops.

DOMESTIC PRODUCTION

U.S. production of petroleum reached its highest level in 1970 at 11.3 million barrels per day total. (See Table 2.2.) Of that amount, 9.6 million barrels per day

TABLE 2.1

Refinery capacity and utilization, 1949–2000

Year	Operable Refineries			Utilization[3] (percent)
	Number[1]	Capacity[2] (million barrels per day)	Gross Input to Distillation Units (million barrels per day)	
1949	336	6.23	5.56	89.2
1950	320	6.22	5.98	92.5
1951	325	6.70	6.76	97.5
1952	327	7.16	6.93	93.8
1953	315	7.62	7.26	93.1
1954	308	7.98	7.27	88.8
1955	296	8.39	7.82	92.2
1956	317	8.58	8.25	93.5
1957	317	9.07	8.22	89.2
1958	315	9.36	8.02	83.9
1959	313	9.76	8.36	85.2
1960	309	9.84	8.44	85.1
1961	309	10.00	8.57	85.7
1962	309	10.01	8.83	88.2
1963	304	10.01	9.14	90.0
1964	298	10.31	9.28	89.6
1965	293	10.42	9.56	91.8
1966	280	10.39	9.99	94.9
1967	276	10.66	10.39	94.4
1968	282	11.35	10.89	94.5
1969	279	11.70	11.25	94.8
1970	276	12.02	11.52	92.6
1971	272	12.86	11.88	90.9
1972	274	13.29	12.43	92.3
1973	268	13.64	13.15	93.9
1974	273	14.36	12.69	86.6
1975	279	14.96	12.90	85.5
1976	276	15.24	13.88	87.8
1977	282	16.40	14.98	89.6
1978	296	17.05	15.07	87.4
1979	308	17.44	14.96	84.4
1980	319	17.99	13.80	75.4
1981	324	18.62	12.75	68.6
1982	301	17.89	12.17	69.9
1983	258	16.86	11.95	71.7
1984	247	16.14	12.22	76.2
1985	223	15.66	12.17	77.6
1986	216	15.46	12.83	82.9
1987	219	15.57	13.00	83.1
1988	213	15.92	13.45	84.7
1989	204	15.65	13.55	86.6
1990	205	15.57	13.61	87.1
1991	202	15.68	13.51	86.0
1992	199	15.70	13.60	87.9
1993	187	15.12	13.85	91.5
1994	179	15.03	14.03	92.6
1995	175	15.43	14.12	92.0
1996	170	15.33	14.34	94.1
1997	164	15.45	14.84	95.2
1998	163	15.71	15.11	95.6
1999	159	16.26	R15.08	R92.6
2000P	158	16.51	15.31	92.6

[1]Prior to 1956, the number of refineries included only those in operation on January 1. For 1957 forward, the number of refineries has included all operable refineries on January 1. See Glossary.
[2]Capacity in million barrels per calendar day on January 1.
[3]For 1949-1980, utilization is derived by dividing gross input to distillation units by one-half of the current year January 1 capacity and the following year January 1 capacity. Percentages were derived from unrounded numbers. For 1981 forward, utilization is derived by averaging reported monthly utilization.
R=Revised. P=Preliminary.
Web Page: http://www.eia.doe.gov/oil_gas/petroleum/pet_frame.html.

SOURCE: "Table 5.9 Refinery Capacity and Utilization, 1949–2000," in *Annual Energy Review 2000*, U.S. Department of Energy, Energy Information Administration, Washington, DC, 2001

was crude oil. (See Figure 2.4.) After 1970 domestic production of petroleum declined. By 2000 U.S. domestic production of petroleum averaged about 8.1 million barrels per day. (See Table 2.2.) Of that amount, 5.8 million barrels per day was crude oil. (See Figure 2.4.) Fig-

ure 2.5 shows the overall flow of oil in the United States for 2000.

In 2000 the 534,000 producing wells in the United States produced an average of 10.9 barrels per day per

TABLE 2.2

Petroleum overview, 1949–2000

(million barrels per day)

Year	Field Production: Crude Oil[1]	Field Production: Natural Gas Plant Liquids	Field Production: Other Liquids[2]	Field Production: Total	Other Domestic Supply[3]	Domestic Supply Total	Imports: Crude Oil[4]	Imports: Products[5]	Imports: Total	Exports	Net Imports	Crude Oil Losses	Petroleum Stock Changes[6]	Products Supplied
1949	5.05	0.43	(7)	5.48	(s)[8]	5.48	0.42	0.22	0.65	0.33	0.32	0.04	-0.01	5.76
1950	5.41	0.50	(7)	5.91	(s)[8]	5.91	0.49	0.36	0.85	0.30	0.55	0.05	-0.06	6.46
1951	6.16	0.56	(7)	6.72	0.01[8]	6.73	0.49	0.35	0.84	0.42	0.42	0.03	0.10	7.02
1952	6.26	0.61	(7)	6.87	0.01[8]	6.88	0.57	0.38	0.95	0.43	0.52	0.02	0.10	7.27
1953	6.46	0.65	(7)	7.11	0.02[8]	7.13	0.65	0.39	1.03	0.40	0.63	0.02	0.14	7.60
1954	6.34	0.69	(7)	7.03	0.02[8]	7.06	0.66	0.40	1.05	0.36	0.70	0.03	-0.03	7.76
1955	6.81	0.77	(7)	7.58	0.02[8]	7.61	0.78	0.47	1.25	0.37	0.88	0.04	(s)	8.46
1956	7.15	0.80	(7)	7.95	0.04[8]	8.00	0.93	0.50	1.44	0.43	1.01	0.05	0.18	8.78
1957	7.17	0.81	(7)	7.98	0.04[8]	8.02	1.02	0.55	1.57	0.57	1.01	0.05	0.17	8.81
1958	6.71	0.81	(7)	7.52	0.06[8]	7.58	0.95	0.75	1.70	0.28	1.42	0.03	-0.14	9.12
1959	7.05	0.88	(7)	7.93	0.09[8]	8.02	0.97	0.81	1.78	0.21	1.57	0.01	0.05	9.53
1960	7.04	0.93	(7)	7.96	0.15[8]	8.11	1.02	0.80	1.81	0.20	1.61	0.01	-0.08	9.80
1961	7.18	0.99	(7)	8.17	0.18[8]	8.35	1.05	0.87	1.92	0.17	1.74	0.01	0.11	9.98
1962	7.33	1.02	(7)	8.35	0.18[8]	8.53	1.13	0.96	2.08	0.17	1.91	0.01	0.03	10.40
1963	7.54	1.10	(7)	8.64	0.20[8]	8.84	1.13	0.99	2.12	0.21	1.91	0.01	(s)	10.74
1964	7.61	1.15	(7)	8.77	0.22[8]	8.99	1.20	1.06	2.26	0.20	2.06	0.01	0.01	11.02
1965	7.80	1.21	(7)	9.01	0.22[8]	9.23	1.24	1.23	2.47	0.19	2.28	0.01	-0.01	11.51
1966	8.30	1.28	(7)	9.58	0.25[8]	9.82	1.22	1.35	2.57	0.20	2.37	0.01	0.10	12.08
1967	8.81	1.41	(7)	10.22	0.29[8]	10.51	1.13	1.41	2.54	0.31	2.23	0.01	0.17	12.56
1968	9.10	1.50	(7)	10.60	0.35[8]	10.95	1.29	1.55	2.84	0.23	2.61	0.01	0.15	13.39
1969	9.24	1.59	(7)	10.83	0.34[8]	11.17	1.41	1.76	3.17	0.23	2.93	0.01	-0.05	14.14
1970	9.64	1.66	(7)	11.30	0.35[8]	11.65	1.32	2.10	3.42	0.26	3.16	0.01	0.10	14.70
1971	9.46	1.69	(7)	11.16	0.44[8]	11.59	1.68	2.25	3.93	0.22	3.70	0.01	0.07	15.21
1972	9.44	1.74	(7)	11.18	0.44[8]	11.63	2.22	2.53	4.74	0.22	4.52	0.01	-0.23	16.37
1973	9.21	1.74	0.03	10.98	0.46	11.43	3.24	3.01	6.26	0.23	6.02	0.01	0.14	17.31
1974	8.77	1.69	0.04	10.50	0.46	10.95	3.48	2.64	6.11	0.22	5.89	0.01	0.18	16.65
1975	8.37	1.63	0.04	10.05	0.48	10.52	4.10	1.95	6.06	0.21	5.85	0.01	0.03	16.32
1976	8.13	1.60	0.04	9.77	0.55	10.33	5.29	2.03	7.31	0.22	7.09	0.01	-0.06	17.46
1977	8.24	1.62	0.05	9.91	0.52	10.43	6.61	2.19	8.81	0.24	8.56	0.02	0.55	18.43
1978	8.71	1.57	0.05	10.33	0.44	10.77	6.36	2.01	8.36	0.36	8.00	0.02	-0.09	18.85
1979	8.55	1.58	0.04	10.18	0.52	10.70	6.52	1.94	8.46	0.47	7.99	0.02	0.17	18.51
1980	8.60	1.57	0.04	10.21	0.63	10.85	5.26	1.65	6.91	0.54	6.36	0.01	0.14	17.06
1981	8.57	1.61	0.05	10.23	0.59	10.82	4.40	1.60	6.00	0.59	5.40	(s)	0.16	16.06
1982	8.65	1.55	0.05	10.25	0.60	10.85	3.49	1.63	5.11	0.82	4.30	(s)	-0.15	15.30
1983	8.69	1.56	0.05	10.30	0.60	10.90	3.33	1.72	5.05	0.74	4.31	(s)	-0.02	15.23
1984	8.88	1.63	0.05	10.55	0.74	11.29	3.43	2.01	5.44	0.72	4.72	(s)	0.28	15.73
1985	8.97	1.61	0.06	10.64	0.70	11.34	3.20	1.87	5.07	0.78	4.29	(s)	-0.10	15.73
1986	8.68	1.55	0.06	10.29	0.76	11.04	4.18	2.05	6.22	0.78	5.44	(s)	0.20	16.28
1987	8.35	1.60	0.06	10.01	0.78	10.79	4.67	2.00	6.68	0.76	5.91	(s)	0.04	16.67
1988	8.14	1.62	0.05	9.82	0.85	10.67	5.11	2.30	7.40	0.82	6.59	(s)	-0.03	17.28
1989	7.61	1.55	0.06	9.22	0.86	10.08	5.84	2.22	8.06	0.86	7.20	(s)	-0.04	17.33
1990	7.36	1.56	0.08	8.99	0.94	9.94	5.89	2.12	8.02	0.86	7.16	(s)	0.11	16.99
1991	7.42	1.66	0.09	9.17	0.91	10.08	5.78	1.84	7.63	1.00	6.63	(s)	-0.01	16.71
1992	7.17	1.70	0.13	9.00	1.03	10.03	6.08	1.80	7.89	0.95	6.94	(s)	-0.07	17.03
1993	6.85	1.74	0.25	8.84	0.93	9.77	6.79	1.83	8.62	1.00	7.62	(s)	0.15	17.24

TABLE 2.2

Petroleum overview, 1949–2000 [CONTINUED]

(million barrels per day)

Year	Domestic Supply						Trade					Crude Oil Losses	Petroleum Stock Changes[6]	Products Supplied
	Field Production				Other Domestic Supply[3]	Total	Imports			Exports	Net Imports			
	Crude Oil[1]	Natural Gas Plant Liquids	Other Liquids[2]	Total			Crude Oil[4]	Products[5]	Total					
1994	6.66	1.73	0.26	8.64	1.03	9.68	7.06	1.93	9.00	0.94	8.05	(s)	0.02	17.72
1995	6.56	1.76	0.30	8.63	0.97	9.59	7.23	1.61	8.83	0.95	7.89	(s)	-0.25	17.72
1996	6.46	1.83	0.31	8.61	1.05	9.66	7.51	1.97	9.48	0.98	8.50	(s)	-0.15	18.31
1997	6.45	1.82	0.34	8.61	0.99	9.61	8.23	1.94	10.16	1.00	9.16	0.00	0.14	18.62
1998	6.25	1.76	0.38	8.39	1.00	9.39	8.71	2.00	10.71	0.94	9.76	(s)	0.24	18.92
1999	R5.88	R1.85	0.38	8.11	1.08	9.18	R8.73	R2.12	R10.85	0.94	R9.91	(s)	R-0.42	R19.52
2000P	5.83	1.91	0.39	8.13	1.25	9.39	8.93	2.16	11.09	1.04	10.05	0.00	-0.04	19.48

[1] Includes lease condensate.
[2] Other hydrocarbons, hydrogen, oxygenates (ethers and alcohols), gasoline blending components, and finished petroleum products.
[3] For 1973 forward, refinery processing gains and unaccounted-for crude oil.
[4] Includes any imports for the Strategic Petroleum Reserve, which began in 1977.
[5] For 1981 forward, includes motor gasoline blending components and aviation gasoline components.
[6] A negative value indicates a net decrease in stocks; a positive value indicates a net increase in stocks. Distillate stocks in the "Northeast Heating Oil Reserve" (2 million barrels at the end of 2000) are not included.
[7] Included in "Other Domestic Supply."
[8] Includes "Other Liquids."
R=Revised. P=Preliminary. (s)=Less than 0.005 million barrels per day and greater than -0.005 million barrels per day.
Notes: Totals may not equal sum of components due to independent rounding. Accurate calculation of the quantity of petroleum products supplied to the domestic market is complicated by the recycling of products at the refinery, the renaming of products involved in a transfer, and the receipt of products from outside the primary supply system. Beginning in 1981, a single adjustment (always a negative quantity) is make to total product supplied to correct this accounting problem. The calculation of this adjustment, called "reclassified," involves only unfinished oils and gasoline blending components. It is the sum of their net changes in primary stocks (net withdrawals is a plus quantity; net additions is a minus quantity) plus imports minus net input to refineries.
Total petroleum products supplied is the sum of the products supplied for each petroleum product, crude oil, unfinished oils, and gasoline blending components. For each of these, except crude oil, product supplied is calculated by adding refinery production, natural gas plant liquids production, new supply of other liquids, imports, and stock withdrawls, and subtracting stock additions, refinery inputs, and exports. Crude oil product supplied is the sum of crude oil burned on leases and at pipeline pump stations as reported on Form EIA-813. Prior to 1983, crude oil burned on leases and at pipeline pump stations was reported as either distillate or residual fuel oil and was included as product supplied for these products. Petroleum product supplied is an approximation of petroleum consumption and is synonymous with the term "Petroleum Consumption" in Section 1. Sector data for petroleum products used in more than one sector are drived from surveys of sales to ultimate consumers by refiners, marketers, distributors, and dealers and from receipts at electric utilities.
Beginning in January 1981, several Energy Information Administration survey forms and calculation methodologies were changed to reflect new developments in refinery and blending plant practices and to improve data integrity. Those changes affect production and product supplied statistics for motor gasoline, distillate fuel oil, and residual fuel oil, and stocks of motor gasoline. On the basis of those changes, motor gasoline production during the last half of 1980 would have averaged, respectively, 105,000 and 54,000 barrels per day higher than the numbers that were published.
Web Page: http://www.eia.doe.gov/oil_gas/petroleum/pet_frame.html.

SOURCE: "Table 5.1 Petroleum Overview, 1949–2000 (Million Barrels per Day)," in *Annual Energy Review 2000*, U.S. Department of Energy, Energy Information Administration, Washington, DC, 2001

FIGURE 2.5

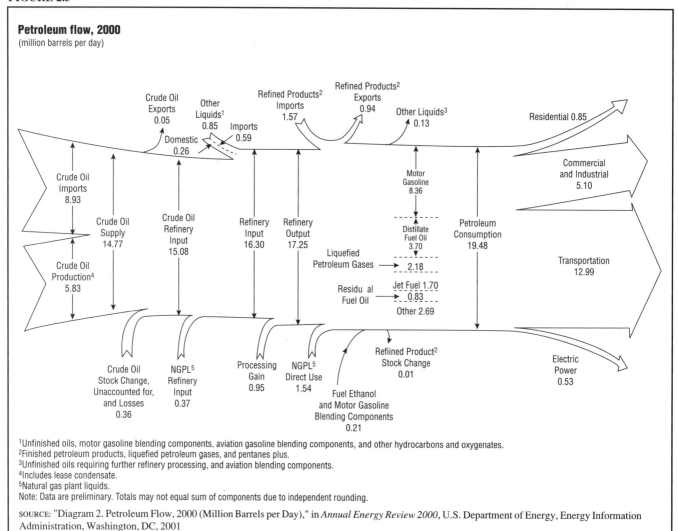

Petroleum flow, 2000
(million barrels per day)

[1] Unfinished oils, motor gasoline blending components, aviation gasoline blending components, and other hydrocarbons and oxygenates.
[2] Finished petroleum products, liquefied petroleum gases, and pentanes plus.
[3] Unfinished oils requiring further refinery processing, and aviation blending components.
[4] Includes lease condensate.
[5] Natural gas plant liquids.
Note: Data are preliminary. Totals may not equal sum of components due to independent rounding.

SOURCE: "Diagram 2. Petroleum Flow, 2000 (Million Barrels per Day)," in *Annual Energy Review 2000*, U.S. Department of Energy, Energy Information Administration, Washington, DC, 2001

well, significantly below peak levels of over 18 barrels per day per well in the early 1970s. Any new oil discoveries are unlikely to lead to a significant increase in domestic production in the near future because of the long lead-time needed to prepare for production.

Most domestic oil production takes place in only a few U.S. states. Texas, Alaska, Louisiana, California, and the offshore areas around these states produce about 75 percent of the nation's oil. Most domestic oil (about 67 percent, or 4 million barrels per day) comes from onshore drilling, while the remaining 2 million barrels come from offshore sources. (See Figure 2.6.) Supplies from Alaska, which increased with the construction of a direct pipeline in the late 1970s, have begun to decline as supplies are used up. (See Figure 2.7.)

Even with oil prices expected to increase in the future, U.S. oil production will likely continue to drop. The Alaskan oil boom appears to be near an end, unless protected wildlife refuges are opened up for drilling. The George W. Bush administration strongly supports a proposal to drill for oil in Alaska's Arctic National

Wildlife Refuge. However, there is great controversy over this proposal, and as of October 2002 no agreement had been reached.

The United States is considered to be in a "mature" oil development phase, meaning that much of the nation's oil has already been found and the nation is producing what it can. The amount of oil discovered per foot of exploratory well in the United States has fallen to less than half the rate of the early 1970s. Of the country's 13 largest oil fields, 7 are at least 80 percent depleted. Geological studies have estimated that 34 percent of U.S. undiscovered recoverable resources are in Alaska, but it is uncertain whether the oil will ever be recovered.

Domestic production is also hampered by the expense of drilling and recovering oil in the United States compared to the expense incurred in Middle Eastern countries. Middle Eastern producers can drill and bring out crude oil from enormous, easily accessible reservoirs for around $2 a barrel. In contrast, the U.S. Department of Energy (DOE) estimates it costs an American oil producer about $14 to produce a barrel of oil, not counting royalty pay-

FIGURE 2.6

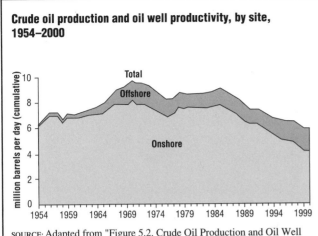

Crude oil production and oil well productivity, by site, 1954–2000

SOURCE: Adapted from "Figure 5.2. Crude Oil Production and Oil Well Productivity, 1954–2000: By Site," in *Annual Energy Review 2000*, U.S. Department of Energy, Energy Information Administration, Washington, DC, 2001. EIA data sources include various U.S. government reports as well as *The Oil Producing Industry in Your State,* Petroleum Association of America, and *World Oil,* Gulf Publishing Co.

FIGURE 2.7

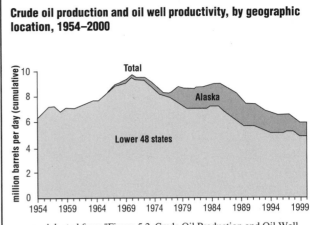

Crude oil production and oil well productivity, by geographic location, 1954–2000

SOURCE: Adapted from "Figure 5.2. Crude Oil Production and Oil Well Productivity, 1954–2000," in *Annual Energy Review 2000*, U.S. Department of Energy, Energy Information Administration, Washington, DC, 2001. EIA data sources include various U.S. government reports as well as *The Oil Producing Industry in Your State,* Petroleum Association of America, and *World Oil,* Gulf Publishing Co.

ments and taxes, which add to the cost. Of all the successful domestic oil wells drilled, only about 1 percent have been "wildcat" wells that have led to the discovery of new large fields, and these discoveries have provided only minor additions to the total proved reserves.

DOMESTIC CONSUMPTION

In 2000 most petroleum, by far, was used for transportation (67 percent), followed by commercial and industrial use (26 percent), residential use (4 percent), and electric utilities (3 percent). (See Figure 2.8.)

Most petroleum used for transportation is for motor gasoline. In the residential and commercial sectors, distillate fuel oil (refined fuels used for space heaters, diesel engines, and electric power generation) accounts for most petroleum use. Liquid petroleum gas (LPG) is the primary oil used in the industrial sector. In electric utilities, residual fuel oils are used most.

A modest decline in residual fuel oil consumption has been caused by the conversion of electric utilities and plants from heavy oil to coal or natural gas energy. An initial decline in the amount of motor gasoline used, beginning in 1978, was attributed to the federal Corporate Average Fuel Economy (CAFE) regulations, which required increased miles-per-gallon efficiency in new automobiles. However, motor gasoline use has increased steadily since then, partly from an increase in users and partly from a leveling off in vehicle efficiency as consumers once again prefer less efficient vehicles, such as larger automobiles and sport utility vehicles (SUVs).

OIL IMPORTS AND EXPORTS

Around the world, there are inequities between those countries that use petroleum and those that possess it.

FIGURE 2.8

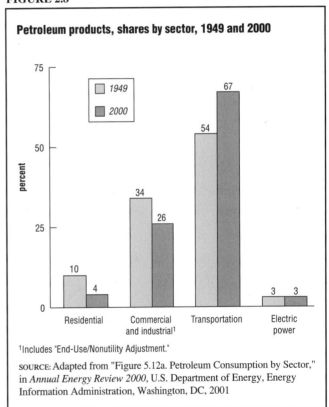

Petroleum products, shares by sector, 1949 and 2000

[1] Includes "End-Use/Nonutility Adjustment."

SOURCE: Adapted from "Figure 5.12a. Petroleum Consumption by Sector," in *Annual Energy Review 2000*, U.S. Department of Energy, Energy Information Administration, Washington, DC, 2001

Countries with surpluses (Saudi Arabia, for example) sell their excess to others that need more than they can produce (the United States, Japan, and European countries). Petroleum is sold as crude oil or as refined products. World trade has been moving toward refined products, as the petroleum-exporting countries have realized that they can make more profit from refined oil products than from crude oil.

Though the United States produces a significant amount of petroleum, it has been importing oil since World War II (1939–45). This reflects the gradual exhaustion of reserves in the United States and the growing energy demand caused by population growth and economic expansion. The relatively low price of foreign oil has encouraged dependence on imported oil. American industry and economic life have been built on oil's cheap availability.

Relatively low crude oil prices and the resulting reduced domestic oil production are the major causes of an increase in imports since 1985. From a low total net import (imports minus exports) of 4.3 million barrels per day in 1985, oil net imports increased by 2000 to 10 million barrels per day. (See Table 2.2.) In 1985 imported oil supplied only 27.3 percent of American oil consumption. Just five years later, in 1990, the proportion had risen to 42 percent, and by 2000 it was 51.6 percent as demand continued to grow. (See Figure 1.7 in Chapter 1.) The Persian Gulf nations (including Saudi Arabia), followed by Canada, Venezuela, and Mexico, were the leading suppliers of petroleum to the United States in 2000.

OIL PRICES

The law of supply and demand usually explains oil price changes; the price of goods reflects a relationship between the supply (availability) and the demand (need). Higher prices increase production, as it becomes profitable to operate more expensive wells, and reduce demand, as consumers lower usage and increase conservation efforts. The factors also work the other way: Reduced demand or increased supply generally cause the price of oil to drop.

While consumers prefer low prices that allow them to save money or get more for the same price, producers naturally prefer to keep prices high. Oil producers formed the OPEC cartel in 1960. A cartel is a group of businesses that agree to control production and marketing to avoid competing with one another. Since 1973 OPEC has tried to control the oil supply in order to achieve higher prices.

OPEC has faced long-term problems, however, because high prices in the late 1970s to mid-1980s encouraged conservation, reducing demand for oil and leading to a sharp decline in oil prices. As a result of the decreased demand for oil and lower prices, OPEC lost some of its ability to control its members and, consequently, prices.

A limited supply of oil was an issue during the Persian Gulf crisis in 1990–91. The conflict caused a sharp, momentary spike in prices as Iraqi and Kuwaiti oil was withdrawn from the market while consumer demand was increasing. The price then dropped as other oil producers, particularly Saudi Arabia, increased production, and an economic recession dampened demand.

Nevertheless, OPEC actions can still effectively influence the petroleum market. In an attempt to halt the downward slide of oil prices in 1999, Saudi Arabia, Mexico, and Venezuela agreed to cut production by 1.6 to 2 million barrels a day. Many other oil-producing nations also limited their production. The limitations worked: During the summer of 2000, oil prices climbed. According to a 2001 report of the Energy Information Administration (EIA) of the DOE (*Annual Energy Review 2000*), the price in real dollars paid by refiners for crude oil in 1999 averaged $16.71 per barrel, and in 2000 averaged $26.40 per barrel.

Gasoline Prices

Many middle-aged Americans can remember when gas cost 30 cents per gallon in the early 1970s. From 1973 to 1981, the price of a gallon of leaded regular gasoline (in current dollars that do not consider inflation) more than tripled, while the price in real dollars (adjusted for inflation) rose 81 percent. (See Table 2.3.) However, after 1981, as a result of the international oil glut, real prices tumbled. In real dollars, the price of a gallon of regular unleaded gasoline was $2.21 in 1981. In 1998 the price was only $1.03, and after price increases in 1999, the price of gas still only averaged $1.41 in 2000.

LESS CONCERN ABOUT OIL DEPENDENCY

In the 1970s the United States and its leaders were very concerned that so much of the U.S. economic structure, based heavily on oil, was dependent upon the whims of OPEC countries. Oil resources became an issue of national security, and OPEC countries, especially the Arab members, were often portrayed as potentially strangling the United States. With about 60 percent of the nation's crude oil supply coming from outside the country, as shown in Figure 2.5, and 46 percent of that coming from OPEC nations, there seems to be little public concern. The Ronald Reagan and George H. W. Bush administrations' decisions to permit the energy issue to be handled by the marketplace was consistent with their economic philosophy but indicated that they saw oil supply as an economic, not a political, issue. This downplayed the international political side of the energy problem. The Clinton administration, as well, was unable to do much about America's dependence on foreign oil, as low prices throughout most of the 1990s set back energy conservation measures and public concern. In the United States, efficiency gains in automobiles have been offset by the preference for large vehicles, such as SUVs. Such preferences imply lack of public interest in reducing the consumption of foreign oil.

The decline in public concern about American's dependence on foreign oil is the result of factors other than low prices and the desire for large vehicles. One factor is a "comfort level" with oil that has been achieved through oil reserves, non-OPEC suppliers, and decreased demand. The United States and European nations have developed substantial reserves to withstand oil supply

TABLE 2.3

Retail motor gasoline and on-highway diesel fuel prices, 1949–2000

(dollars per gallon)

Year	Motor gasoline by grade[1]								Regular motor gasoline by area type[2,3]					On-highway diesel fuel[3]
	Leaded regular		Unleaded regular		Unleaded premium		All types		Conventional	Oxygenated	Oxygenated and reformulated	Reformulated	All area types	
	Nominal	Real[4]	Nominal	Real[4]	Nominal	Real[4]	Nominal	Real[4]						
1949	0.27	1.55	NA	NA	NA	NA	NA	NA	NA	NA	NA	NA	NA	NA
1950	0.27	1.54	NA	NA	NA	NA	NA	NA	NA	NA	NA	NA	NA	NA
1951	0.27	1.45	NA	NA	NA	NA	NA	NA	NA	NA	NA	NA	NA	NA
1952	0.27	1.44	NA	NA	NA	NA	NA	NA	NA	NA	NA	NA	NA	NA
1953	0.29	1.49	NA	NA	NA	NA	NA	NA	NA	NA	NA	NA	NA	NA
1954	0.29	1.49	NA	NA	NA	NA	NA	NA	NA	NA	NA	NA	NA	NA
1955	0.29	1.47	NA	NA	NA	NA	NA	NA	NA	NA	NA	NA	NA	NA
1956	0.30	1.46	NA	NA	NA	NA	NA	NA	NA	NA	NA	NA	NA	NA
1957	0.31	1.47	NA	NA	NA	NA	NA	NA	NA	NA	NA	NA	NA	NA
1958	0.30	1.41	NA	NA	NA	NA	NA	NA	NA	NA	NA	NA	NA	NA
1959	0.31	1.39	NA	NA	NA	NA	NA	NA	NA	NA	NA	NA	NA	NA
1960	0.31	1.40	NA	NA	NA	NA	NA	NA	NA	NA	NA	NA	NA	NA
1961	0.31	1.37	NA	NA	NA	NA	NA	NA	NA	NA	NA	NA	NA	NA
1962	0.31	1.35	NA	NA	NA	NA	NA	NA	NA	NA	NA	NA	NA	NA
1963	0.30	1.32	NA	NA	NA	NA	NA	NA	NA	NA	NA	NA	NA	NA
1964	0.30	1.30	NA	NA	NA	NA	NA	NA	NA	NA	NA	NA	NA	NA
1965	0.31	1.31	NA	NA	NA	NA	NA	NA	NA	NA	NA	NA	NA	NA
1966	0.32	1.31	NA	NA	NA	NA	NA	NA	NA	NA	NA	NA	NA	NA
1967	0.33	1.32	NA	NA	NA	NA	NA	NA	NA	NA	NA	NA	NA	NA
1968	0.34	1.28	NA	NA	NA	NA	NA	NA	NA	NA	NA	NA	NA	NA
1969	0.35	1.26	NA	NA	NA	NA	NA	NA	NA	NA	NA	NA	NA	NA
1970	0.36	1.23	NA	NA	NA	NA	NA	NA	NA	NA	NA	NA	NA	NA
1971	0.36	1.19	NA	NA	NA	NA	NA	NA	NA	NA	NA	NA	NA	NA
1972	0.36	1.14	NA	NA	NA	NA	NA	NA	NA	NA	NA	NA	NA	NA
1973	0.39	1.16	NA	NA	NA	NA	NA	NA	NA	NA	NA	NA	NA	NA
1974	0.53	1.45	NA	NA	NA	NA	NA	NA	NA	NA	NA	NA	NA	NA
1975	0.57	1.42	NA	NA	NA	NA	NA	NA	NA	NA	NA	NA	NA	NA
1976	0.59	1.40	0.61	1.45	NA	NA	NA	NA	NA	NA	NA	NA	NA	NA
1977	0.62	1.38	0.66	1.46	NA	NA	NA	NA	NA	NA	NA	NA	NA	NA
1978	0.63	1.30	0.67	1.39	NA	NA	0.65	1.35	NA	NA	NA	NA	NA	NA
1979	0.86	1.64	0.90	1.73	NA	NA	0.88	1.69	NA	NA	NA	NA	NA	NA
1980	1.19	2.09	1.25	2.18	NA	NA	1.22	2.14	NA	NA	NA	NA	NA	NA
1981	1.31	2.10	1.38	2.21	1.47	2.36	1.35	2.17	NA	NA	NA	NA	NA	NA
1982	1.22	1.85	1.30	1.96	1.42	2.14	1.28	1.93	NA	NA	NA	NA	NA	NA
1983	1.16	1.68	1.24	1.80	1.38	2.01	1.23	1.78	NA	NA	NA	NA	NA	NA
1984	1.13	1.58	1.21	1.70	1.37	1.91	1.20	1.68	NA	NA	NA	NA	NA	NA
1985	1.12	1.51	1.20	1.63	1.34	1.82	1.20	1.62	NA	NA	NA	NA	NA	NA
1986	0.86	1.14	0.93	1.23	1.09	1.44	0.93	1.24	NA	NA	NA	NA	NA	NA
1987	0.90	1.16	0.95	1.22	1.09	1.41	0.96	1.23	NA	NA	NA	NA	NA	NA
1988	0.90	1.12	0.95	1.18	1.11	1.38	0.96	1.20	NA	NA	NA	NA	NA	NA

TABLE 2.3

Retail motor gasoline and on-highway diesel fuel prices, 1949–2000 [CONTINUED]

(dollars per gallon)

Year	Motor gasoline by grade[1] Leaded regular Nominal	Leaded regular Real[4]	Unleaded regular Nominal	Unleaded regular Real[4]	Unleaded premium Nominal	Unleaded premium Real[4]	All types Nominal	All types Real[4]	Regular motor gasoline by area type[2,3] Conventional	Oxygenated	Oxygenated and reformulated	Reformulated	All area types	On-highway diesel fuel[3]
1989	1.00	1.20	1.02	1.23	1.20	1.44	1.06	1.27	NA	NA	NA	NA	NA	NA
1990	1.15	1.33	1.16	1.35	1.35	1.56	1.22	1.41	NA	NA	NA	NA	NA	NA
1991	NA	NA	1.14	1.27	1.32	1.47	1.20	1.33	1.10	NA	NA	NA	1.10	NA
1992	NA	NA	1.13	1.23	1.32	1.43	1.19	1.30	1.09	NA	NA	NA	1.09	NA
1993	NA	NA	1.11	1.18	1.30	1.38	1.17	1.25	1.05	1.14	NA	NA	1.07	NA
1994	NA	NA	1.11	1.16	1.31	1.36	1.17	1.22	1.06	1.14	NA	NA	1.08	NA
1995	NA	NA	1.15	1.17	1.34	1.36	1.21	1.23	1.09	1.16	1.18	1.16	1.11	1.11
1996	NA	NA	1.23	1.23	1.41	1.41	1.29	1.29	1.18	1.27	1.27	1.24	1.20	1.24
1997	NA	NA	1.23	1.21	1.42	1.39	1.29	1.27	1.18	1.26	1.28	1.25	1.20	1.20
1998	NA	NA	1.06	1.03	1.25	1.21	1.12	1.08	1.01	1.08	1.09	1.08	1.03	1.04
1999	NA	NA	1.17	1.11	1.36	1.30	1.22	1.17	1.11	1.20	1.19	1.20	1.14	1.12
2000	NA	NA	1.51	1.41	1.69	1.58	1.56	1.46	1.46	1.51	1.55	1.54	1.48	1.49

[1]Average motor gasoline prices are calculated from a sample of service stations providing all types of service (i.e., full-, mini-, and self-serve). Geographic coverage - 1949-1973, 55 representative cities; 1974-1977, 56 urban areas; 1978 forward, 85 urban areas.

[2]"Area Type" refers to the specific types of motor gasoline that are mandated by the Environmental Protection Agency to be sold in designated areas of the country. Only cash self-service prices are included.

[3]Nominal dollars.

[4]In chained (1996) dollars, calculated by using gross domestic product implicit price deflators.

NA=Not available.

SOURCE: "Table 5.22. Retail Motor Gasoline and On-Highway Diesel Fuel Prices, 1949–2000," in *Annual Energy Review 2000*, U.S. Department of Energy, Energy Information Administration, Washington, DC, 2001. EIA data sources include *Platt's Oil Price Handbook and Oilmanac, 1974*, and various U.S. government reports.

fluctuations. These reserves include the Strategic Petroleum Reserve (SPR) in the United States and government-required reserves in Europe. Furthermore, in emergencies, non-OPEC oil producers such as the United Kingdom and Norway can increase their output. Demand for oil has stabilized because of the conservation efforts of many industrialized nations, particularly European nations.

The increased use of pipelines across Saudi Arabia and Turkey has made the job of picking up oil from these countries safer. A growing number of tankers pick up their oil in either the Red Sea or the Mediterranean Sea before delivering it to Europe or the United States. These ships do not have to go through the potentially dangerous Persian Gulf. In addition, since many American strategists are wary of navigation in the Straits of Hormuz, where a future enemy might be able to stop the flow of oil to the West, shipment through pipelines lessens the importance of the waterway. On the other hand, such pipelines could be destroyed relatively easily in a war.

Based on these factors, if the oil shortages that developed in 1973 or 1979 occurred again, the result would likely not be the same. Despite the U.S. role as protector of Kuwait's oil in the Persian Gulf, there has seemed to be little serious concern that America's dependence on foreign oil, especially OPEC oil, represents a threat to national security or national stability. This may change based on the terrorist attacks of September 11, 2001, and their aftermath, the "War on Terrorism." Experts suggest that if the United States attacks Iraq, as the Bush administration was considering in late 2002, the ultimate result might be a stabilization of the supply and price of oil (Michael E. Kanell, "War in Iraq: Is It About Oil?" *The [Montreal] Gazette,* October 26, 2002).

Oil Dependency and the Persian Gulf War

The 1990–91 Persian Gulf War illustrated the effectiveness of both the reserves and market mechanisms. Petroleum prices rose in reaction to the loss of Iraqi and Kuwaiti oil supplies and speculation that hostilities in the region could restrict other Persian Gulf oil supplies. Demand dropped in reaction to the higher prices, which in turn was one factor in the recession that followed. In addition, the price increases were somewhat moderated by the availability of reserve supplies, as the United States and other Western nations agreed to release reserves. The United States even sold some of its oil to other countries. Even though the World Bank predicted that oil prices could go as high as $65 a barrel in 1991, that scenario did not occur. (See Figure 1.2 in Chapter 1 for historical oil price fluctuations.)

After the moderate price increases in 1990–91, prices fell. Saudi Arabia and other producers, taking advantage of the war-reduced supply and higher prices, had increased production. Production continued to rise even as demand fell, resulting in record-high world petroleum stockpiles. As the imbalance persisted, oil prices fell to below the precrisis level. Persian Gulf oil production remained high after the war. (See Figure 1.2 in Chapter 1.)

Nonetheless, the war highlighted the dependence on oil by the United States and its Western allies. The United States and its allies committed 500,000 troops to war. Though world leaders spoke of stopping aggression, many observers wondered whether the reaction would have been the same if Kuwait did not have large oil fields—that is, whether the United States entered the war in large part to guarantee its oil supply.

CONSERVATION AND OIL DEMAND

The relatively slow growth in total demand for oil can be attributed to conservation measures adopted in response to dramatic oil price increases in the 1970s and early 1980s. Energy efficiency is increasing as oil-powered equipment continues to be replaced with new, more efficient models. Some Americans have switched from oil to other forms of fuel, including renewable resources, to heat their homes. Moderate investments in insulation and heater efficiency in older buildings have been able to cut fuel needs for these buildings by 25 percent.

Most of the increase in oil demand has been for transportation. Much more efficient vehicles can be built with today's technology, but low gasoline prices have eliminated much of the consumer demand for them.

Because the United States is a "mature" oil-producing nation, with few new sources of domestic oil likely to be located, an increased national commitment to conservation could be an important way to resolve the problem of an ever-greater dependence on foreign oil. Some observers believe that it is better to develop programs promoting conservation now, because the cost might be less than if steps had to be taken in a hurried manner during a future crisis. Others argue, however, that the marketplace will resolve the problem if shortages arise. Either way, the once important issue of "energy independence" is no longer considered feasible, but diversification of energy sources has become the present-day goal.

ENVIRONMENTAL CONCERNS ABOUT OIL TRANSPORTATION

Fossil fuels, including petroleum, have been responsible for much contamination of the environment. Because of public sentiment and legislation passed to slow environmental damage, efforts are being made to shift to other means of energy production, to conserve existing oil supplies, and to find cleaner ways to produce, transport, and burn oil.

Transporting oil carries significant environmental risks. According to the U.S. Department of the Interior, the cause of most transportation spills is oil tanker accidents, such as the grounding of the *Exxon Valdez* in 1989.

The *Exxon Valdez* Oil Spill

A number of events have influenced American attitudes toward oil production and use. In one of the most notable, from March 1989, the *Exxon Valdez* oil tanker hit a reef in Alaska, spilling 11 million gallons of crude oil into the waters of Prince William Sound. The cleanup cost Exxon $1.28 billion, which does not include legal costs or any valuation of the wildlife lost. The spill was an environmental disaster for a formerly pristine area. Even measures used to clean up the spill, such as washing the beaches with hot water, caused additional damage.

The *Exxon Valdez* spill also led to debate about added safety measures in the design of tankers. Tankers are bigger than ever before. In 1945 the largest tanker held 16,500 tons of oil; today, supertankers carry more than 550,000 tons. These supertankers are difficult to maneuver because of their size and are likely to spill more oil if damaged. Although there have been fewer spills since 1973, the amount of oil lost has been roughly the same.

The Oil Pollution Act of 1990

The *Exxon Valdez* oil spill led Congress to pass the Oil Pollution Act of 1990 (PL 101-380) after having debated the issue for 16 years. The bill increases, but still limits, oil spillers' federal liability (financial responsibility) as long as a spill is not the result of "gross negligence." The bill also mandates compensation to those who are economically injured by oil-spill accidents. Damages that can be charged to oil companies are limited to $60 million for tanker accidents and $75 million for accidents at offshore facilities. The rest of the cleanup costs are paid from a $500 million oil-spill fund generated by a 1.3 cents-per-barrel tax on oil. Individual states still maintain the right to impose unlimited liability on spillers. Oil companies are also required to phase in double hulls on oil vessels over a 25-year period by 2015. Double hulls provide an extra container in case of accidents.

Accidents Still Occur Worldwide

In mid-November 2002, the 26-year-old, single-hulled tanker *Prestige* was first damaged, then sank in the Atlantic Ocean off the coast of Spain. The Spanish government estimates the ship spilled an estimated 5,000 tons of heavy fuel oil when it was damaged and an additional 12,000 tons as it sank. Environmentalists believe these figures may be higher. It is believed that much of the 77,000 tons of oil the tanker originally contained remained in the sunken hull at the bottom of the Atlantic as of early December 2002. If this entire amount of oil spills into the ocean, the *Prestige* could leak a total of 20.6 million gallons of oil, nearly double the amount spilled from the *Exxon Valdez*. According to a news story posted on cnn.com on December 5, 2002 ("Spanish Military Battle Oil Spill"), it was believed the sunken tanker was still leaking oil.

The oil spilled from the *Prestige* has caused the deaths of unknown numbers of fish, dolphins, and

FIGURE 2.9

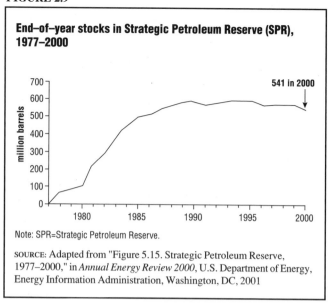

End–of–year stocks in Strategic Petroleum Reserve (SPR), 1977–2000

Note: SPR=Strategic Petroleum Reserve.

SOURCE: Adapted from "Figure 5.15. Strategic Petroleum Reserve, 1977–2000," in *Annual Energy Review 2000*, U.S. Department of Energy, Energy Information Administration, Washington, DC, 2001

seabirds and is responsible for both economic and environmental damage to the Galician fish and shellfish industries. As of early December 2002, about 556 miles of Spanish coastline had been closed to fishing, and as a result, over 4,000 fishermen were out of work.

THE SHRINKING U.S. OIL INDUSTRY

During the past two decades, the U.S. petroleum industry has experienced a severe loss of jobs. The number of seismic land crews and marine vessels searching for oil in the United States and its waters decreased sharply after 1981. From 1982 to 1992 alone, oil-extraction companies lost 51 percent of their workforce, petroleum refiners, 28 percent. In Texas, once the center of the U.S. oil industry, jobs in the industry plummeted from 80,000 in 1981 to 25,000 in 1996. In 1998 the Texas Comptroller of Public Accounts estimated that for every $1 drop in the price of oil per barrel, 10,000 jobs are lost in the Texas economy. That translated into 100,000 jobs lost in the Texas oil industry from October 1997 to December 1998. According to a February 6, 1999 article in the *New York Times* ("Oil Industry Sees More Losses"), the oil industry lost 24,415 jobs between November 1997 and February 1999 because of the decline of oil prices during that period. Oil companies blamed environmental regulations and the restriction of oil exploration on some federal lands, such as Arctic wildlife preserves and some offshore areas. Additionally, mergers in the industry cost thousands of jobs as companies restructured and downsized. Job cuts in the oil industry extended into 2002.

STRATEGIC PETROLEUM RESERVES

In 1923 the Harding administration set up the Petroleum Reserve to ensure that the U.S. Navy would have ade-

FIGURE 2.10

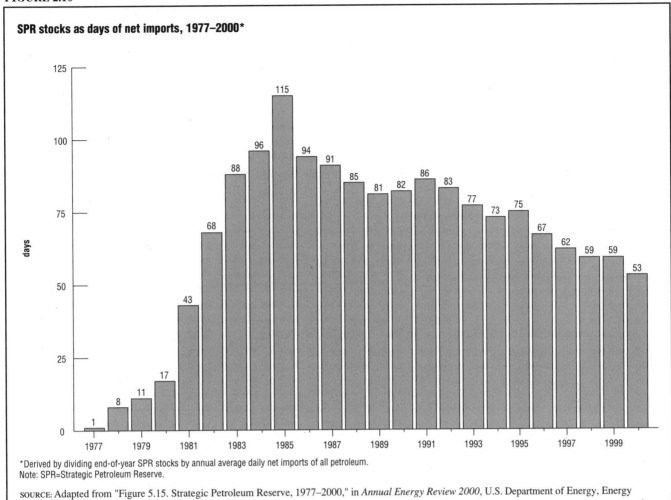

SPR stocks as days of net imports, 1977–2000*

*Derived by dividing end-of-year SPR stocks by annual average daily net imports of all petroleum.
Note: SPR=Strategic Petroleum Reserve.

SOURCE: Adapted from "Figure 5.15. Strategic Petroleum Reserve, 1977–2000," in *Annual Energy Review 2000*, U.S. Department of Energy, Energy Information Administration, Washington, DC, 2001

quate fuel in the event of war. In 1975, in response to the growing concern over America's energy dependence, Congress turned the Strategic Petroleum Reserve (SPR) over to the Department of the Interior under the Energy Policy and Conservation Act (PL 94-163). An additional law, PL 101-383, expanded the SPR and created a second reserve for refined products. A small amount of oil was withdrawn from the reserves at the start of the Persian Gulf War, but large withdrawals were unnecessary.

In the SPR oil is stored in deep salt caverns located at four storage sites in Louisiana and two in Texas. (Oil does not dissolve salt the way water does.) Each site's caverns vary in capacity, but most can hold approximately 10 million barrels of oil. The SPR system currently has 41 such caverns. If the United States suddenly found its supplies cut off, the reserve system would be connected to existing commercial lines that would start pumping the oil.

At the end of 2000 the SPR contained 541 million barrels (see Figure 2.9), enough to equal 53 days of imported oil should the supply be cut off (see Figure 2.10). The International Energy Agency calculated that

the United States held 123 days of emergency reserves at that time. Approximately 43 percent of reserves (541 million barrels) was held by the government, while the rest (57 percent or 717 million barrels) was held privately. Figure 2.10 shows a decline in the reserves, from a high of 115 days in 1985 to the current 53 days. This decline is the result of two major factors:

1. It reflects the increase in imports since 1985. As the nation has imported a greater amount of oil, the days of import replacement represented by the amount of oil in the SPR have dropped.

2. It reflects a real decline in quantity. In the mid-1990s, millions of barrels of oil were sold (at a loss of $10 a barrel or so) to help balance the federal budget. In September 2000 President Clinton authorized the release of 30 million barrels of oil to bolster oil supplies, particularly heating oil in the Northeast.

SYNTHETIC FUELS

Fuel can be produced from other energy sources besides oil, including coal, natural gas, grain, garbage,

FIGURE 2.11

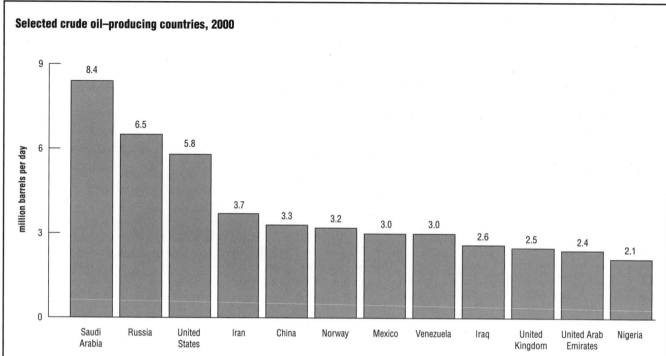

Selected crude oil–producing countries, 2000

million barrels per day

Country	Value
Saudi Arabia	8.4
Russia	6.5
United States	5.8
Iran	3.7
China	3.3
Norway	3.2
Mexico	3.0
Venezuela	3.0
Iraq	2.6
United Kingdom	2.5
United Arab Emirates	2.4
Nigeria	2.1

SOURCE: Adapted from "Figure 11.4. World Crude Oil Production," in *Annual Energy Review 2000*, U.S. Department of Energy, Energy Information Administration, Washington, DC, 2001. EIA data sources include various U.S. government reports as well as information obtained from the CIA; *Narodnoye Khozyaystvo SSSR*, USSR Central Statistical Office; and *Annual Statistical Bulletin*, Organization of Petroleum Exporting Countries.

and shale oil. (Shale oil is obtained from bituminous shale, a type of rock containing tarlike substances.) Some common examples of such synthetic (human-made) fuels are alcohol, methane gas, methanol, and gasohol. Many years ago, kerosene was made from coal oil; it was an early synthetic fuel.

In 1980 President Jimmy Carter signed legislation creating the Synthetic Fuels Corporation Act (PL 96-294) in an attempt to make the United States less dependent on foreign oil. The legislation was intended to provide funding for companies that would turn coal, shale oil, and tar sands into oil and natural gas. Congress originally financed the program with $15 billion, acting on consumer fears that the price of oil had no upper limit and expert predictions that oil prices would reach $90 per barrel. The synthetic fuels industry quickly lost support when the United States experienced a recession in the 1980s and world oil prices dropped, making synthetic fuels uneconomical. It cost much less to import oil than to manufacture synthetic fuels, and the idea of energy independence quickly faded away. In its five-year existence, the Synthetic Fuels Corporation planned six manufacturing plants. After gradual reductions in financing, the entire program was discontinued in 1985.

The U.S. government has continued a program to study and subsidize the use of ethanol, which is grain alcohol usually produced from corn, as a gasoline extender or substitute. Ethanol receives subsidies in the form of fuel

tax exemptions that compensate for its higher cost. Additionally, state governments provide incentives to ethanol producers by paying them a set fee per gallon of ethanol they produce. An ethanol-gasoline mixture is usually 10 percent ethanol and 90 percent gasoline. Gasoline with higher percentages of ethanol tends to corrode parts of automobile fuel systems, makes cold engines hard to start, and produces less energy per gallon, which reduces fuel efficiency. Brazil pursues an aggressive program to substitute fuels from grain for gasoline, with mixed results.

INTERNATIONAL OIL PRODUCTION AND CONSUMPTION

World Production

Total world petroleum production generally increased after the early 1980s, reaching nearly 68 million barrels per day in 2000. (See Table 2.4.) The major producers in 2000 were Saudi Arabia, Russia, and the United States. (See Figure 2.11.) Together, these three countries accounted for 30 percent of the world's crude oil production. Other leading oil producers include Iran, China, Norway, Mexico, Venezuela, Iraq, the United Kingdom, United Arab Emirates, and Nigeria.

World Consumption

World petroleum consumption in 1999 was 74.9 million barrels per day. (See Table 2.5.) The United States was by far the leading consumer, using 19.5 million barrels per

TABLE 2.4

World crude oil production, 1960–2000
(million barrels per day)

	Selected OPEC[1] producers									Selected non-OPEC producers									
Year	Persian Gulf Nations[2]	Iran	Iraq	Kuwait[3]	Nigeria	Saudi Arabia[3]	United Arab Emirates	Venezuela	Total OPEC	Canada	China	Mexico	Norway	Former U.S.S.R.	Russia	United Kingdom	United States	Total non-OPEC[4]	World
1960	5.27	1.07	0.97	1.69	0.02	1.31	0.00	2.85	8.70	0.52	0.10	0.27	0.00	2.91	—	(s)	7.04	12.29	20.99
1961	5.65	1.20	1.01	1.74	0.05	1.48	0.00	2.92	9.36	0.61	0.11	0.29	0.00	3.28	—	(s)	7.18	13.09	22.45
1962	6.19	1.33	1.01	1.96	0.07	1.64	0.01	3.20	10.51	0.67	0.12	0.31	0.00	3.67	—	(s)	7.33	13.84	24.35
1963	6.82	1.49	1.16	2.10	0.08	1.79	0.05	3.25	11.51	0.71	0.13	0.31	0.00	4.07	—	(s)	7.54	14.62	26.13
1964	7.61	1.71	1.26	2.30	0.12	1.90	0.19	3.39	12.98	0.75	0.18	0.32	0.00	4.60	—	(s)	7.61	15.20	28.18
1965	8.37	1.91	1.32	2.36	0.27	2.21	0.28	3.47	14.35	0.81	0.23	0.33	0.00	4.79	—	(s)	7.80	15.98	30.33
1966	9.32	2.13	1.39	2.48	0.42	2.60	0.36	3.37	15.77	0.88	0.29	0.36	0.00	5.23	—	(s)	8.30	17.19	32.96
1967	9.91	2.60	1.23	2.50	0.32	2.81	0.38	3.54	16.85	0.96	0.28	0.36	0.00	5.68	—	(s)	8.81	18.54	35.39
1968	10.91	2.84	1.50	2.61	0.14	3.04	0.50	3.60	18.79	1.19	0.30	0.39	0.00	6.08	—	(s)	9.10	19.84	38.63
1969	11.95	3.38	1.52	2.77	0.54	3.22	0.63	3.59	20.91	1.13	0.48	0.46	0.00	6.48	—	(s)	9.24	20.79	41.70
1970	13.39	3.83	1.55	2.99	1.08	3.80	0.78	3.71	23.30	1.26	0.60	0.49	0.00	6.99	—	(s)	9.64	22.59	45.89
1971	15.77	4.54	1.69	3.20	1.53	4.77	1.06	3.55	25.21	1.35	0.78	0.49	0.01	7.48	—	(s)	9.46	23.31	48.52
1972	17.54	5.02	1.47	3.28	1.82	6.02	1.20	3.22	26.89	1.53	0.90	0.51	0.03	7.89	—	(s)	9.44	24.25	51.14
1973	20.67	5.86	2.02	3.02	2.05	7.60	1.53	3.37	30.63	1.80	1.09	0.47	0.03	8.32	—	(s)	9.21	25.05	55.68
1974	21.28	6.02	1.97	2.55	2.26	8.48	1.68	2.98	30.35	1.55	1.32	0.57	0.04	8.91	—	(s)	8.77	25.37	55.72
1975	18.93	5.35	2.26	2.08	1.78	7.08	1.66	2.35	26.77	1.43	1.49	0.71	0.19	9.52	—	0.01	8.37	26.06	52.83
1976	21.51	5.88	2.42	2.15	2.07	8.58	1.94	2.29	30.33	1.31	1.67	0.83	0.28	10.06	—	0.25	8.13	27.01	57.34
1977	21.73	5.66	2.35	1.97	2.09	9.25	2.00	2.24	30.89	1.32	1.87	0.98	0.28	10.60	—	0.77	8.24	28.82	59.71
1978	20.61	5.24	2.56	2.13	1.90	8.30	1.83	2.17	29.46	1.32	2.08	1.21	0.36	11.11	—	1.08	8.71	30.70	60.16
1979	21.07	3.17	3.48	2.50	2.30	9.53	1.83	2.36	30.58	1.50	2.12	1.46	0.40	11.38	—	1.57	8.55	32.09	62.67
1980	17.96	1.66	2.51	1.66	2.06	9.90	1.71	2.17	26.61	1.44	2.11	1.94	0.53	11.71	—	1.62	8.60	32.99	59.60
1981	15.25	1.38	1.00	1.13	1.43	9.82	1.47	2.10	22.48	1.29	2.01	2.31	0.50	11.85	—	1.81	8.57	33.60	56.08
1982	12.16	2.21	1.01	0.82	1.30	6.48	1.25	1.90	18.78	1.27	2.05	2.75	0.52	11.91	—	2.07	8.65	34.70	53.48
1983	11.08	2.44	1.06	1.06	1.24	5.09	1.15	1.80	17.50	1.36	2.12	2.69	0.61	11.97	—	2.29	8.69	35.76	53.26
1984	10.78	2.17	1.21	1.16	1.39	4.66	1.15	1.80	17.44	1.44	2.30	2.78	0.70	11.86	—	2.48	8.88	37.05	54.49
1985	9.63	2.25	1.43	1.02	1.50	3.39	1.19	1.68	16.18	1.47	2.51	2.75	0.79	11.59	—	2.53	8.97	37.80	53.98
1986	11.70	2.04	1.69	1.42	1.47	4.87	1.33	1.79	18.28	1.47	2.62	2.44	0.87	11.90	—	2.54	8.68	37.95	56.23
1987	12.10	2.30	2.08	1.59	1.34	4.27	1.54	1.75	18.52	1.54	2.69	2.55	1.02	12.05	—	2.41	8.35	38.15	56.67
1988	13.46	2.24	2.69	1.49	1.45	5.09	1.57	1.90	20.32	1.62	2.73	2.51	1.16	12.05	—	2.23	8.14	38.42	58.74
1989	14.84	2.81	2.90	1.78	1.72	5.06	1.86	1.91	22.07	1.56	2.76	2.52	1.55	11.72	—	1.80	7.61	37.79	59.86

TABLE 2.4

World crude oil production, 1960–2000 [CONTINUED]

(million barrels per day)

| Year | Selected OPEC[1] producers | | | | | | | | | Selected non-OPEC producers | | | | | | | | | |
	Persian Gulf Nations[2]	Iran	Iraq	Kuwait[3]	Nigeria	Saudi Arabia[3]	United Arab Emirates	Venezuela	Total OPEC	Canada	China	Mexico	Norway	Former U.S.S.R.	Russia	United Kingdom	United States	Total non-OPEC[4]	World
1990	15.28	3.09	2.04	1.18	1.81	6.41	2.12	2.14	23.20	1.55	2.77	2.55	1.70	10.98	—	1.82	7.36	37.37	60.57
1991	14.74	3.31	0.31	0.19	1.89	8.12	2.39	2.38	23.27	1.55	2.84	2.68	1.89	9.99	—	1.80	7.42	36.94	60.21
1992	15.97	3.43	0.43	1.06	1.94	8.33	2.27	2.37	24.40	1.61	2.85	2.67	2.23	—	7.63	1.83	7.17	35.81	60.21
1993	16.71	3.54	0.51	1.85	1.96	8.20	2.16	2.45	25.12	1.68	2.89	2.67	2.35	—	6.73	1.92	6.85	35.12	60.24
1994	16.96	3.62	0.55	2.03	1.93	8.12	2.19	2.59	25.51	1.75	2.94	2.69	2.52	—	6.14	2.37	6.66	35.48	60.99
1995	17.21	3.64	0.56	2.06	1.99	8.23	2.23	2.75	26.00	1.81	2.99	2.62	2.77	—	6.00	2.49	6.56	36.33	62.33
1996	17.37	3.69	0.58	2.06	2.00	8.22	2.28	2.94	26.46	1.84	3.13	2.86	3.10	—	5.85	2.57	6.46	37.25	63.71
1997	R18.10	3.66	1.16	R2.01	R2.13	R8.36	2.32	R3.28	R27.71	1.92	3.20	3.02	3.14	—	5.92	2.52	6.45	R37.98	R65.69
1998	R19.34	3.63	2.15	2.09	2.15	8.39	2.35	3.17	R28.77	1.98	3.20	3.07	3.02	—	R5.85	2.62	6.25	R38.19	R66.96
1999	R18.67	3.56	2.51	1.90	2.13	7.83	2.17	2.83	R27.58	1.91	R3.19	2.91	3.02	—	R6.08	R2.68	R5.88	R38.29	R65.87
2000P	19.94	3.72	2.57	2.13	2.14	8.40	2.35	2.95	29.11	1.98	3.25	3.01	3.20	—	6.48	2.47	5.83	38.87	67.98

[1]Organization of Petroleum Exporting Countries. (OPEC)—Algeria, Indonesia, Iran, Iraq, Kuwait, Libya, Nigeria, Qatar, Sudi Arabia, United Arab Emirates, and Venezuela.
[2]Persian Gulf Nations are Bahrain, Iran, Iraq, Kuwait, Qatar, Saudi Arabia, and United Arab Emirates.
[3]Includes about one-half of the production in the Neutral Zone between Kuwait and Saudi Arabia.
[4]Ecuador, which withdrew from OPEC on December 31, 1992, and Gabon, which withdrew on December 31, 1994, are included in "Non-OPEC" for all years.
R=Revised. P=Preliminary. — = Not applicable. (s)=Less than 0.005 million barrels per day.
Note: Includes lease condensate, excludes natural gas plant liquids. Totals may not equal sum of components due to independent rounding.

SOURCE: "Table 11.4 World Crude Oil Production, 1960–2000," in *Annual Energy Review 2000*, U.S. Department of Energy, Energy Information Administration, Washington, DC, 2001. EIA data sources include various U.S. government reports as well as information obtained from the CIA; *Narodnoye Khozyaystvo SSSR*, USSR Central Statistical Office; and *Annual Statistical Bulletin*, Organization of Petroleum Exporting Countries.

TABLE 2.5

World petroleum consumption, 1960–1999

(million barrels per day)

Year	Selected OECD¹ consumers											Selected non-OECD consumers						
	Canada	France	Germany²	Italy	Japan	Mexico³	South Korea³	Spain	United Kingdom	United States	Total OECD⁴	Brazil	China	India	Former U.S.S.R.	Russia	Total non-OECD	World
1960	0.84	0.56	0.63	0.44	0.66	0.30	0.01	0.10	0.94	9.80	15.78	0.27	0.17	0.16	2.38	—	5.56	21.34
1961	0.87	0.63	0.79	0.54	0.82	0.29	0.02	0.12	1.04	9.98	16.77	0.28	0.17	0.17	2.57	—	6.23	23.00
1962	0.92	0.73	1.00	0.67	0.93	0.30	0.02	0.12	1.12	10.40	18.06	0.31	0.14	0.18	2.87	—	6.83	24.89
1963	0.99	0.86	1.17	0.77	1.21	0.31	0.03	0.12	1.27	10.74	19.60	0.34	0.17	0.21	3.15	—	7.32	26.92
1964	1.05	0.98	1.36	0.90	1.48	0.33	0.02	0.20	1.36	11.02	21.05	0.35	0.20	0.22	3.58	—	8.03	29.08
1965	1.14	1.09	1.61	0.98	1.74	0.34	0.03	0.23	1.49	11.51	22.81	0.33	0.23	0.25	3.61	—	8.33	31.14
1966	1.21	1.19	1.80	1.08	1.98	0.36	0.04	0.31	1.58	12.08	24.60	0.38	0.30	0.28	3.87	—	8.96	33.56
1967	1.25	1.34	1.86	1.19	2.14	0.39	0.07	0.36	1.64	12.56	25.94	0.38	0.28	0.26	4.22	—	9.65	35.59
1968	1.34	1.46	1.99	1.40	2.66	0.41	0.10	0.46	1.82	13.39	28.56	0.46	0.31	0.31	4.48	—	10.40	38.96
1969	1.42	1.66	2.33	1.69	3.25	0.45	0.15	0.49	1.98	14.14	31.54	0.48	0.44	0.34	4.87	—	11.35	42.89
1970	1.52	1.94	2.83	1.71	3.82	0.50	0.20	0.58	2.10	14.70	34.49	0.53	0.62	0.40	5.31	—	12.32	46.81
1971	1.56	2.12	2.94	1.84	4.14	0.52	0.23	0.64	2.14	15.21	36.07	0.58	0.79	0.42	5.66	—	13.35	49.42
1972	1.66	2.32	3.13	1.95	4.36	0.59	0.23	0.68	2.28	16.37	38.74	0.66	0.91	0.46	6.12	—	14.35	53.09
1973	1.73	2.60	3.34	2.07	4.95	0.67	0.28	0.78	2.34	17.31	41.53	0.78	1.12	0.49	6.60	—	15.71	57.24
1974	1.78	2.45	3.06	2.00	4.86	0.71	0.29	0.86	2.21	16.65	40.12	0.86	1.19	0.47	7.28	—	16.56	56.68
1975	1.78	2.25	2.96	1.86	4.62	0.75	0.31	0.87	1.91	16.32	38.82	0.92	1.36	0.50	7.52	—	17.38	56.20
1976	1.82	2.42	3.21	1.97	4.84	0.83	0.36	0.97	1.89	17.46	41.39	1.00	1.53	0.51	7.78	—	18.28	59.67
1977	1.85	2.29	3.21	1.90	4.88	0.88	0.42	0.94	1.91	18.43	42.43	1.02	1.64	0.55	8.18	—	19.40	61.83
1978	1.90	2.41	3.29	1.95	4.95	0.99	0.48	0.98	1.94	18.85	43.62	1.11	1.79	0.62	8.48	—	20.54	64.16
1979	1.97	2.46	3.37	2.04	5.05	1.10	0.53	1.02	1.97	18.51	44.01	1.18	1.84	0.66	8.64	—	21.21	65.22
1980	1.87	2.26	3.08	1.93	4.96	1.27	0.54	0.99	1.73	17.06	41.41	1.15	1.77	0.64	9.00	—	21.66	63.07
1981	1.77	2.02	2.80	1.87	4.85	1.40	0.54	0.94	1.59	16.06	39.14	1.09	1.71	0.73	8.94	—	21.76	60.90
1982	1.58	1.88	2.74	1.78	4.58	1.48	0.53	1.00	1.59	15.30	37.45	1.06	1.66	0.74	9.08	—	22.05	59.50
1983	1.45	1.84	2.66	1.75	4.40	1.35	0.56	1.01	1.53	15.23	36.59	0.98	1.73	0.77	8.95	—	22.15	58.74
1984	1.47	1.75	2.66	1.65	4.58	1.45	0.59	0.91	1.85	15.73	37.43	1.03	1.74	0.82	8.91	—	22.41	59.84
1985	1.50	1.78	2.70	1.72	4.38	1.47	0.57	0.85	1.63	15.73	37.23	1.08	1.89	0.90	8.95	—	22.87	60.10
1986	1.51	1.77	2.86	1.74	4.44	1.49	0.61	0.88	1.65	16.28	38.28	1.24	2.00	0.95	8.98	—	23.48	61.76
1987	1.55	1.79	2.77	1.86	4.48	1.52	0.64	0.90	1.60	16.67	38.96	1.26	2.12	0.99	9.00	—	24.04	63.00
1988	1.69	1.80	2.74	1.84	4.75	1.55	0.73	0.98	1.70	17.28	40.24	1.30	2.28	1.08	8.89	—	24.58	64.82
1989	1.73	1.86	2.58	1.93	4.98	1.64	0.84	1.03	1.74	17.33	40.88	1.32	2.38	1.15	8.74	—	25.04	65.92

TABLE 2.5

World petroleum consumption, 1960–1999 [CONTINUED]

(million barrels per day)

	Selected OECD[1] consumers											Selected non-OECD consumers						
Year	Canada	France	Germany[2]	Italy	Japan	Mexico[3]	South Korea[3]	Spain	United Kingdom	United States	Total OECD[4]	Brazil	China	India	Former U.S.S.R.	Russia	Total non-OECD	World
1990	1.69	1.82	2.66	1.87	5.14	1.68	1.03	1.01	1.75	16.99	40.92	1.34	2.30	1.17	8.39	—	R25.05	R65.97
1991	1.62	1.94	2.83	1.86	5.28	1.70	1.20	1.07	1.80	16.71	41.40	1.35	2.50	1.19	8.35	—	R25.16	R66.56
1992	1.64	1.93	2.84	1.94	5.45	1.72	1.46	1.11	1.80	17.03	42.42	1.37	2.66	1.28	—	4.42	24.34	66.76
1993	1.69	1.88	2.90	1.85	5.40	1.71	1.69	1.06	1.82	17.24	42.98	1.43	2.96	1.31	—	3.75	24.02	67.00
1994	1.73	1.83	2.88	1.84	5.67	1.80	1.86	1.13	1.84	17.72	44.17	1.51	3.16	1.41	—	3.18	R24.12	R68.29
1995	1.76	1.90	2.88	2.05	5.71	1.72	2.03	1.26	1.85	17.72	44.96	1.60	3.36	1.58	—	2.98	R24.92	R69.88
1996	1.80	1.94	2.91	2.06	5.87	1.76	2.18	1.18	1.85	18.31	46.07	1.72	3.61	1.68	—	2.62	R25.34	R71.41
1997	1.86	1.96	2.90	2.05	5.71	1.87	2.39	1.30	1.80	18.62	46.83	1.82	3.92	1.77	—	2.56	R26.23	R73.06
1998	R1.86	2.03	2.92	2.07	5.51	R1.94	R1.97	1.39	R1.77	18.92	R46.93	R1.92	4.11	1.84	—	R2.45	R26.71	73.64
1999P	1.93	2.03	2.82	1.98	5.57	1.98	2.04	1.43	1.72	19.52	47.61	1.95	4.32	1.93	—	2.40	27.30	74.91

[1]Organization for Economic Cooperation and Development.—as of December 31, 1999, its members were Australia, Austria, Belgium, Canada, Czech Republic, Denmark, Finland, France, Germany, Greece, Hungary, Iceland, Ireland, Italy Japan, Luxembourg, Mexico, Netherlands, New Zealand, Norway, Poland, Portugal, South Korea, Spain, Sweden, Switzerland, Turkey, United Kingdom, and United States.

[2]Through 1969, the data for Germany are for the former West Germany only. For 1970 through 1990, this is East and West Germany. Beginning in 1991, this is unified Germany.

[3]Mexico, which joined the OECD on May 18, 1994, and South Korea, which joined the OECD on December 12, 1996, are included in the OECD for all years shown in this table.

[4]Hungary and Poland, which joined the OECD on May 7, 1996, and November 22, 1996, respectively, are included in Total OECD beginning in 1970, the first year that data for these countries were available. The Czech Republic, which joined the OECD on December 21, 1995, is included in Total OECD beginning in 1993, the year that it came into existence.

Note: Totals may not equal sum of components due to independent rounding.

R=Revised. P=Preliminary. — = Not applicable.

SOURCE: "Table 11.9. World Petroleum Consumption, 1960–1999," in *Annual Energy Review 2000*, U.S. Department of Energy, Energy Information Administration, Washington, DC, 2001.

day, followed by Japan (5.6 million barrels per day), China (4.3 million barrels a day), Germany (2.8 million barrels a day), and Russia (2.4 million barrels per day).

FUTURE TRENDS IN THE OIL INDUSTRY

The EIA, in its *Annual Energy Outlook 2002* (2001), projected that domestic crude oil production in the lower 48 states will generally decline from 4.9 million barrels per day in 2000 to about 4.5 million barrels per day by 2020. Total offshore production of crude oil was 1.6 million barrels per day in 2000 and is projected to increase to 1.8 million in 2020. Crude oil production in Alaska is expected to decline to 0.7 million barrels per day in 2010, then grow to 1.1 million barrels per day by 2020. In spite of these decreases in domestic crude oil production, the domestic petroleum supply is projected to increase from its 2000 level of 9.3 million barrels per day to 10.0 million barrels per day in 2020. (See Figure 2.12.) This is expected to be the result of an increase in production of natural gas plant liquids and gains in the efficiency of refining oil.

Domestic petroleum consumption is expected to grow at an average annual rate of 1.5 percent through 2020 to approximately 27 million barrels per day, depending on economic growth and world oil prices. More than 70 percent of the increased consumption of petroleum in 2020 will be in the transportation sector. In transportation, effi-

FIGURE 2.12

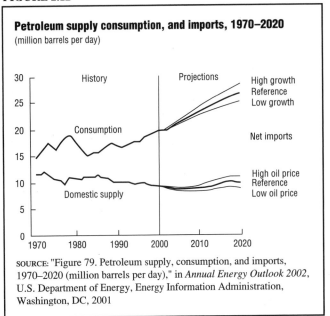

Petroleum supply consumption, and imports, 1970–2020
(million barrels per day)

SOURCE: "Figure 79. Petroleum supply, consumption, and imports, 1970–2020 (million barrels per day)," in *Annual Energy Outlook 2002*, U.S. Department of Energy, Energy Information Administration, Washington, DC, 2001

ciency gains are projected to slow while travel is projected to increase. The demand for light-duty vehicles such as SUVs and for heavy trucks is expected to increase as well. From 2000 to 2010 the projected gap between petroleum consumption and domestic petroleum supply is expected to result in an increase in net imports of petroleum.

CHAPTER 3
NATURAL GAS

Natural gas is an important source of energy in the United States. Methane, ethane, and propane are the primary constituents of natural gas, with methane making up 73 to 95 percent of the total.

The natural gas industry developed out of the petroleum industry. Wells drilled for oil often produced considerable amounts of natural gas, but early oilmen had no idea what to do with it. Originally considered a waste by-product of oil production, natural gas had no market, nor were transmission lines available to deliver it even if a use had been known. As a result, the gas was burned off, or flared. Pictures of southeast Texas in the early twentieth century show thousands of wooden drilling rigs topped with plumes of gas flaming like burning candles. Even today, flaring sites are sometimes the brightest areas visible in nighttime satellite images, outshining even the largest urban areas.

Nonetheless, researchers soon found ways to use natural gas. In 1925 the first natural gas pipeline, over 200 miles long, was built from Louisiana to Texas. U.S. demand grew rapidly, especially after World War II (1939–45). By the 1950s natural gas was providing one-quarter of the nation's energy needs. Today, natural gas is second only to coal in the share of U.S. energy produced, beating out even crude oil, which falls third. (See Table 1.1 in Chapter 1.) A vast pipeline transmission system connects production facilities in the United States, Canada, and Mexico with natural gas distributors. Figure 3.1 shows the production and consumption figures for natural gas for 2000. Figure 3.2 shows the pattern of natural gas supply and distribution in the United States in 2000.

THE PRODUCTION OF NATURAL GAS

Natural gas is produced from gas wells and oil wells. There is little delay between production and consumption, except for some gas, which is placed in storage. Changes in demand are almost immediately reflected by changes in wellhead flows, or supply.

Total natural gas production in America was 19.2 trillion cubic feet in 2000, below the peak levels produced from 1969 to 1975. (See Figure 3.3 and Figure 3.4.) Although production levels are being driven up by increasing demand and rising prices, production continues to be outpaced by consumption. Imported gas makes up the difference between supply and demand. As regards domestic production, Texas, Louisiana, and Oklahoma accounted for over half the natural gas produced in the United States in 2000.

Natural Gas Wells

In 2000, 323,000 gas wells were in operation in the United States, the highest number ever. (See Figure 3.5.) Although the number of producing wells increased steadily after 1960 and more sharply after the mid-1970s, there are slight fluctuations from drilling for new sources of gas, abandoning old wells, weather conditions, and the economic condition of the nation. Wells are abandoned when they are no longer economical to run.

The average productivity of natural gas wells dropped through the mid-1980s after hitting peak productivity in 1971; it has remained at a steady low level since then. (See Figure 3.6.)

Offshore Production

Offshore drilling for natural gas accounted for about one-fifth of the total U.S. production in 2000. (See Figure 3.7.) Almost all natural gas produced offshore comes from the Gulf of Mexico and offshore California. U.S. offshore production is expected to increase to meet the nation's growing need for energy, although this type of production could be slowed by environmental restrictions.

Offshore drilling generally occurs on the outer continental shelf, the submerged area offshore with a depth of up to 200 meters (656 feet). Figure 3.8 is a diagram of a continental margin. The continental shelf varies from one

FIGURE 3.1

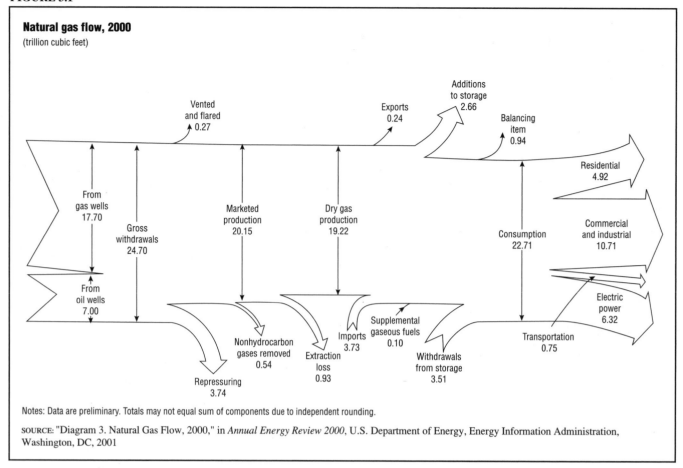

Natural gas flow, 2000
(trillion cubic feet)

Notes: Data are preliminary. Totals may not equal sum of components due to independent rounding.

SOURCE: "Diagram 3. Natural Gas Flow, 2000," in *Annual Energy Review 2000*, U.S. Department of Energy, Energy Information Administration, Washington, DC, 2001

coastal area to another. The shelf is relatively narrow along the Pacific coast, wide along much of the Atlantic coast and the Gulf of Alaska, and widest in the Gulf of Mexico. The U.S. Department of the Interior has leased more than 1.5 billion acres of offshore areas to oil companies for offshore drilling.

The development of offshore oil and gas resources began with the drilling of the Summerland oil field in California in 1896, where about 400 wells were drilled. In the search for oil and gas in offshore areas, the industry has continually improved drilling technology. Today, deepwater petroleum exploration occurs from platforms and drill ships, and shallow-water explorations from gravel islands and mobile units.

Even though natural gas is transported mostly by pipelines instead of by tankers, the 1989 Exxon oil spill in Prince William Sound, Alaska, and other oil spills have focused national attention on all types of offshore drilling. Even before the *Exxon Valdez* oil spill, environmentalists were calling for the curtailment of offshore drilling for both oil and gas.

Natural Gas Reserves

Reserves are estimated volumes of gas in known deposits that are believed to be recoverable in the future.

Proved reserves are those gas volumes that geological and engineering data show with reasonable certainty to be recoverable. Proved reserves of natural gas amounted to about 176 trillion cubic feet in 1999. (See Table 3.1.) Reserves in Texas, the Gulf of Mexico, Oklahoma, and Louisiana make up approximately half the total.

Natural gas reserves in North America are generally more abundant than crude oil reserves, although historically they have been difficult to estimate accurately. At one time the U.S. Department of Energy (DOE) estimated that proven supplies of recoverable gas in the United States would last fewer than eight years. However, new discoveries and technological improvements have increased the estimated recoverable supply of natural gas to 50 years or more.

The North Slope fields of Alaska are estimated to contain reserves amounting to 35 trillion cubic feet, but at the moment there is no easy way to transport Alaska's reserves to the lower 48 states. Low natural gas prices have made building a pipeline from Alaska uneconomical. By 2002, however, a proposal for building an Alaskan natural gas pipeline had been passed by the Senate in the nation's energy bill, but it had yet to pass the House. Lawmakers were still debating the route the pipeline would take and other issues (such as energy subsidies) as of November 2002. If built, the pipeline could deliver 4.5 billion cubic

FIGURE 3.2

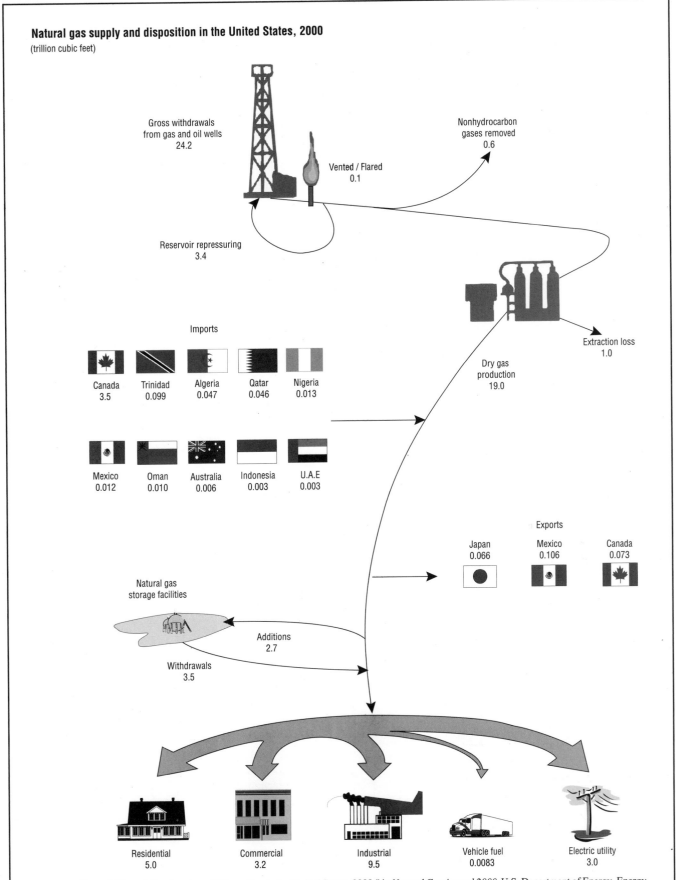

Natural gas supply and disposition in the United States, 2000
(trillion cubic feet)

Gross withdrawals from gas and oil wells 24.2

Nonhydrocarbon gases removed 0.6

Vented / Flared 0.1

Reservoir repressuring 3.4

Extraction loss 1.0

Imports

| Canada 3.5 | Trinidad 0.099 | Algeria 0.047 | Qatar 0.046 | Nigeria 0.013 |
| Mexico 0.012 | Oman 0.010 | Australia 0.006 | Indonesia 0.003 | U.A.E 0.003 |

Dry gas production 19.0

Exports

| Japan 0.066 | Mexico 0.106 | Canada 0.073 |

Natural gas storage facilities

Additions 2.7

Withdrawals 3.5

Residential 5.0

Commercial 3.2

Industrial 9.5

Vehicle fuel 0.0083

Electric utility 3.0

SOURCE: "Figure 3. Natural Gas Supply and Disposition in the United States, 2000," in *Natural Gas Annual 2000*, U.S. Department of Energy, Energy Information Administration, Washington, DC, 2001.

FIGURE 3.3

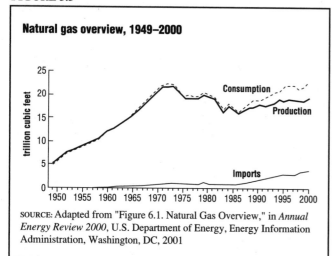

Natural gas overview, 1949–2000

SOURCE: Adapted from "Figure 6.1. Natural Gas Overview," in *Annual Energy Review 2000*, U.S. Department of Energy, Energy Information Administration, Washington, DC, 2001

FIGURE 3.5

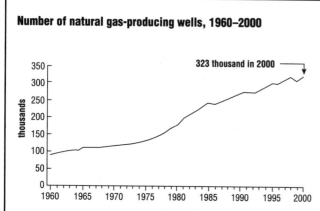

Number of natural gas-producing wells, 1960–2000

323 thousand in 2000

SOURCE: Adapted from "Figure 6.4. Natural Gas Gross Withdrawals by State and Location and Gas Well Productivity, 1960–2000," in *Annual Energy Review 2000*, U.S. Department of Energy, Energy Information Administration, Washington, DC, 2001. EIA data sources include various U.S. government reports as well as *World Oil*, Gulf Publishing Co.

FIGURE 3.7

Natural gas gross withdrawals by location, 1960–2000

SOURCE: Adapted from "Figure 6.4. Natural Gas Gross Withdrawals by State and Location and Gas Well Productivity, 1960–2000," in *Annual Energy Review 2000*, U.S. Department of Energy, Energy Information Administration, Washington, DC, 2001. EIA data sources include various U.S. government reports as well as *World Oil*, Gulf Publishing Co.

FIGURE 3.4

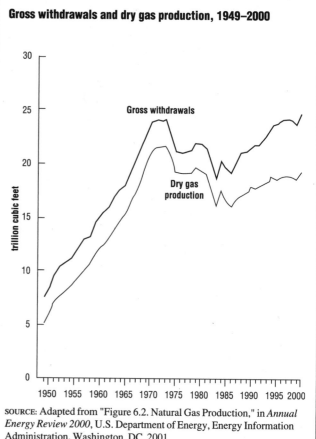

Gross withdrawals and dry gas production, 1949–2000

SOURCE: Adapted from "Figure 6.2. Natural Gas Production," in *Annual Energy Review 2000*, U.S. Department of Energy, Energy Information Administration, Washington, DC, 2001

FIGURE 3.6

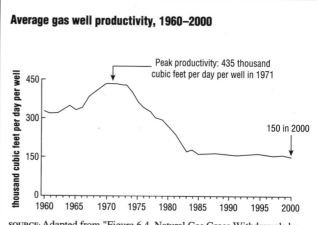

Average gas well productivity, 1960–2000

Peak productivity: 435 thousand cubic feet per day per well in 1971

150 in 2000

SOURCE: Adapted from "Figure 6.4. Natural Gas Gross Withdrawals by State and Location and Gas Well Productivity, 1960–2000," in *Annual Energy Review 2000*, U.S. Department of Energy, Energy Information Administration, Washington, DC, 2001. EIA data sources include various U.S. government reports as well as *World Oil*, Gulf Publishing Co.

feet of natural gas per day to the lower 48 states—about 10 percent of the nation's daily gas consumption.

Underground Storage

Because of seasonal, daily, and even hourly changes in gas demand, substantial natural gas storage facilities have been created to meet supply needs. Many of these storage centers are depleted gas reservoirs located near transmission lines and marketing areas. Gas is injected into storage when market needs are lower than the available gas flow, and gas is withdrawn from storage when supplies from producing fields and/or the capacity of transmission lines are not adequate to meet peak demands. At the end of 2000, gas in underground storage totaled 6.1 trillion cubic feet. (See Figure 3.9.)

TRANSMISSION OF NATURAL GAS

A vast network of natural gas pipelines crisscrosses the United States, connecting every state except Alaska, Hawaii, and Vermont. (Vermont receives its gas directly from Canada and is not connected to the U.S. pipeline.) The natural gas in this quarter-million-mile system generally flows northeastward, primarily from Texas and Louisiana, the two major gas-producing states, and to a lesser extent from Oklahoma and New Mexico. (See Figure 3.10.) It also flows west to California.

Imports enter the United States via pipeline from Canada in Idaho, Maine, Michigan, Montana, New Hampshire, New York, North Dakota, Washington, and Vermont.

It also enters via pipeline in Texas from Mexico. These pipeline imports make up 94 percent of total U.S. imports of natural gas. Liquefied natural gas is shipped to the United States from Algeria, Australia, Indonesia, Nigeria, Oman, Qatar, Trinidad, and the United Arab Emirates.

FIGURE 3.8

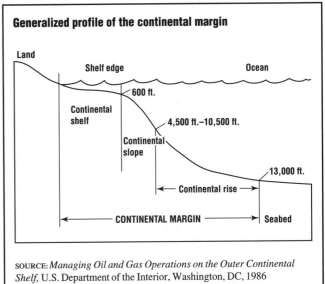

Generalized profile of the continental margin

SOURCE: *Managing Oil and Gas Operations on the Outer Continental Shelf*, U.S. Department of the Interior, Washington, DC, 1986

TABLE 3.1

Crude oil and natural gas field counts, cumulative production, proved reserves, and ultimate recovery, 1977–1999

Year	Cumulative number of fields with crude oil and/or natural gas	Cumulative number of fields with crude oil	Crude oil and lease condensate (billion barrels)			Cumulative number of fields with natural gas	Natural gas[1] (trillion cubic feet)		
			Cumulative production	Proved reserves	Proved ultimate recovery		Cumulative production	Proved reserves	Proved ultimate recovery
1977	31,360	27,835	121.4	33.6	155.0	23,883	558.3	209.5	767.8
1978	32,430	28,683	124.6	33.1	157.6	24,786	578.4	210.1	788.5
1979	33,644	29,671	127.7	31.2	158.9	25,823	599.1	208.3	807.4
1980	34,999	30,766	130.8	31.3	162.2	26,919	619.4	206.3	825.6
1981	36,621	32,111	133.9	31.0	165.0	28,213	639.4	209.4	848.9
1982	38,123	33,375	137.1	29.5	166.6	29,375	658.1	209.3	867.4
1983	39,489	34,495	140.3	29.3	169.6	30,419	675.1	209.0	884.1
1984	41,038	35,784	143.5	30.0	173.5	31,595	693.5	206.0	899.5
1985	42,317	36,849	146.8	29.9	176.7	32,595	710.9	202.2	913.1
1986	43,076	37,464	150.0	28.3	178.3	33,151	727.8	201.1	928.9
1987	43,742	37,982	153.0	28.7	181.7	33,657	745.4	196.4	941.8
1988	44,414	38,506	156.0	28.2	184.2	34,196	763.4	177.0	940.4
1989	44,883	38,858	158.8	27.9	186.7	34,579	781.7	175.4	957.1
1990	45,385	39,244	161.5	27.6	189.0	34,975	800.4	177.6	978.0
1991	45,776	39,558	164.2	25.9	190.1	35,254	819.1	175.3	994.4
1992	46,149	39,843	166.8	25.0	191.8	35,539	838.0	173.3	1,011.3
1993	46,513	40,124	169.3	24.1	193.4	35,798	857.2	170.5	1,027.7
1994	46,922	40,417	171.7	23.6	195.3	36,142	877.1	171.9	1,049.1
1995	47,296	40,694	174.1	23.5	197.7	36,433	896.9	173.5	1,070.4
1996	47,557	40,875	176.5	23.3	199.8	36,612	917.0	175.1	1,092.1
1997	47,854	40,977	178.9	23.9	202.8	36,830	937.1	175.7	1,112.8
1998	[2]47,664	[2]35,143	181.2	R22.4	R203.5	[2]32,458	957.0	R172.4	R1,129.4
1999	NA	NA	183.3	23.2	206.5	NA	976.8	176.2	1,153.0

[1]Wet, after lease separation.
[2]There is a discontinuity in this time series between 1997 and 1998 due to the absence of updates for a subset of the data used in the past.
Note: Data are at end of year

SOURCE: "Table 4.2. Crude Oil and Natural Gas Field Counts, Cumulative Production, Proved Reserves, and Ultimate Recovery, 1977–1999," in *Annual Energy Review 2000*, U.S. Department of Energy, Energy Information Administration, Washington, DC, 2001.

FIGURE 3.9

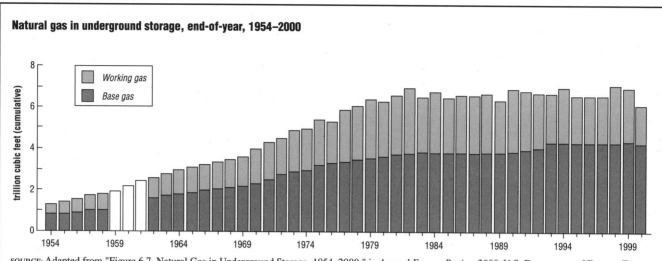

Natural gas in underground storage, end-of-year, 1954–2000

SOURCE: Adapted from "Figure 6.7. Natural Gas in Underground Storage, 1954–2000," in *Annual Energy Review 2000*, U.S. Department of Energy, Energy Information Administration, Washington, DC, 2001. EIA data sources include various U.S. government reports as well as *Gas Facts*, American Gas Association.

FIGURE 3.10

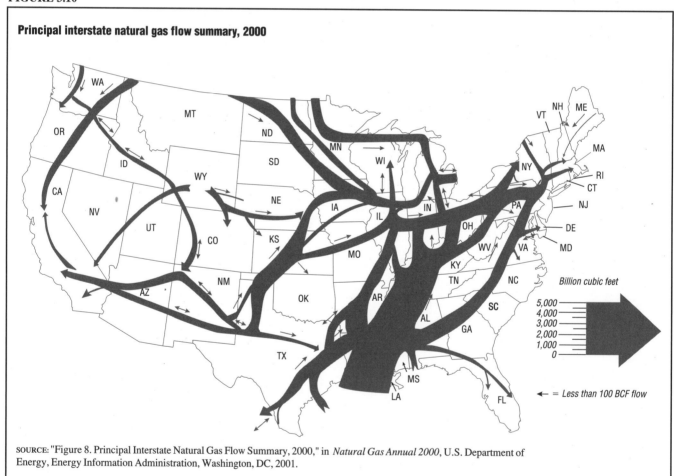

Principal interstate natural gas flow summary, 2000

SOURCE: "Figure 8. Principal Interstate Natural Gas Flow Summary, 2000," in *Natural Gas Annual 2000*, U.S. Department of Energy, Energy Information Administration, Washington, DC, 2001.

The largest users of natural gas in the residential sector in 2000 were California, Illinois, and New York.

DOMESTIC NATURAL GAS CONSUMPTION

Nationally, natural gas consumption has been generally rising since 1986, the low point of demand since 1965. (See Table 3.2.) Natural gas use was 22.7 trillion cubic feet in 2000. Of this amount, 47 percent was used by industry, 22 percent by residences, 15 percent by commercial customers, and 13 percent by electric utilities; 3 percent was used as pipeline fuel in the gas transporting process.

Natural gas fills an important part of the country's energy needs. It is an attractive fuel not only because its price is relatively low but also because it is clean and efficient and can help the country meet both its environmental goals and its energy needs.

The residential sector used 4.9 trillion cubic feet of natural gas in 2000. (See Table 3.2.) Energy consumption by residences depends heavily on weather-related home-heating demands. The colder it gets, the more gas is used. According to the U.S. Department of Commerce, approximately one-half of all residential energy consumers in the United States use gas to heat their homes. Conservation practices and efficiency of gas appliances such as water heaters and stoves also affect residential consumption patterns.

The use of natural gas in the commercial sector was 3.4 trillion cubic feet in 2000. (See Table 3.2.) Like residential consumption, use in the commercial sector depends heavily on seasonal requirements, as well as the number of users and conservation measures taken by commercial establishments. Commercial customers are particularly sensitive to changes in gas prices and in the economy.

The industrial sector has historically been the largest consumer of natural gas. Consumption in this sector in 2000 was at an all-time high of 10.6 trillion cubic feet, surpassing the previous high of 10.2 trillion cubic feet in 1973. (See Table 3.2.) Substitution of natural gas for petroleum for some industrial purposes has increased consumption in the industrial sector.

NATURAL GAS PRICES

Natural gas prices can vary because of differing federal and state rate structures. Region also plays a role—for example, prices are lower in major natural gas–producing areas where transmission costs are lower. From the mid-twentieth century through the early 1970s, natural gas prices were relatively stable. (See Figure 3.11.) Thereafter, deregulation and industry restructuring brought about a period of sharply rising prices, with wellhead prices (the value of natural gas at the mouth of the well) reaching a high in 1983, declining until 1991, and then varying through the rest of the 1990s. The average price of all categories of natural gas at the wellhead in 2000 was $3.37 per 1,000 cubic feet, sharply rising from $2.07 in 1999.

At the retail price level in real dollars, residential customers paid $6.39 per thousand cubic feet of natural gas in 1999 and $7.20 in 2000. (See Table 3.3.) Commercial consumers paid $5.76 per thousand cubic feet in 2000, while industrial consumers paid $4.16 per thousand cubic feet.

Much of the variation in natural gas prices through the years can be attributed to changes in the natural gas industry. The passage of the Natural Gas Policy Act of 1978 (NGPA; PL 95-621) triggered a dramatic transformation in the natural gas industry. The NGPA allowed gas prices at the wellhead to rise gradually. (See Figure 3.11.) On January 1, 1985, new gas prices were decontrolled, and additional volumes of onshore production were decontrolled on July 1, 1987. In 1988 President Ronald Reagan signed legislation removing all remaining natural gas wellhead price controls by 1993.

The NGPA allowed prices to go up, but it also opened the market to the forces of supply and demand. Now that prices are decontrolled and the industry is no longer constrained by federal regulations, the natural gas industry has become more sensitive to market signals and is able to respond more quickly to changes in economic conditions.

NATURAL GAS IMPORTS AND EXPORTS

U.S. natural gas trading was limited to the neighboring countries of Mexico and Canada until shipping natural gas in liquefied form became a feasible alternative to pipelines. In 1969 the first shipments of liquefied natural gas (LNG) were sent from Alaska to Japan, and U.S. imports of LNG from Algeria began the following year.

In 2000 U.S. net imports of natural gas by all routes totaled a record 3.5 trillion cubic feet, approximately 15.4 percent of domestic consumption. (See Figure 3.12.) Natural gas imports have been increasing significantly since 1986. Historically, Canada has been, by far, the major supplier of U.S. natural gas imports, accounting for nearly all of the natural gas imported in 2000.

According to the DOE's Energy Information Administration (EIA) in *Annual Energy Review 2000* (published in 2001), the United States exported 237 billion cubic feet of natural gas in 2000. Of these exports, Japan and Mexico bought the largest amount (174 billion cubic feet), while Canada purchased 63 billion cubic feet.

INTERNATIONAL NATURAL GAS USAGE

World Production

The world's gross production of dry natural gas totaled an all-time high of 84.7 trillion cubic feet in 1999. (See Figure 3.13.) Russia and the United States were by

TABLE 3.2

Natural gas consumption by sector, 1949–2000

(trillion cubic feet)

Year	Residential — Delivered to residences	Commercial[2] — Delivered to commercial facilities	Industrial[2] — Delivered to industrial facilities	Industrial[2] — Lease and plant fuel	Industrial[2] — Total	Transportation — Pipeline fuel[3]	Transportation — Delivered for vehicle fuel use	Transportation — Total	End-Use sectors — Total	Electric power sector[1] — Electric utilities	Electric power sector[1] — Nonutility power producers	Electric power sector[1] — Total	End-use/nonutility adjustment[4]	Total consumption
1949	0.99	0.35	2.25	0.84	3.08	NA	NA	NA	4.42	0.55	NA	NA	—	4.97
1950	1.20	0.39	2.50	0.93	3.43	0.13	NA	0.13	5.14	0.63	NA	NA	—	5.77
1951	1.47	0.46	2.77	1.15	3.91	0.19	NA	0.19	6.05	0.76	NA	NA	—	6.81
1952	1.62	0.52	2.87	1.16	4.04	0.21	NA	0.21	6.38	0.91	NA	NA	—	7.29
1953	1.69	0.53	3.03	1.13	4.16	0.23	NA	0.23	6.60	1.03	NA	NA	—	7.64
1954	1.89	0.58	3.07	1.10	4.17	0.23	NA	0.23	6.88	1.17	NA	NA	—	8.05
1955	2.12	0.63	3.41	1.13	4.54	0.25	NA	0.25	7.54	1.15	NA	NA	—	8.69
1956	2.33	0.72	3.71	1.00	4.71	0.30	NA	0.30	8.05	1.24	NA	NA	—	9.29
1957	2.50	0.78	3.89	1.05	4.93	0.30	NA	0.30	8.51	1.34	NA	NA	—	9.85
1958	2.71	0.87	3.89	1.15	5.03	0.31	NA	0.31	8.93	1.37	NA	NA	—	10.30
1959	2.91	0.98	4.22	1.24	5.46	0.35	NA	0.35	9.69	1.63	NA	NA	—	11.32
1960	3.10	1.02	4.53	1.24	5.77	0.35	NA	0.35	10.24	1.72	NA	NA	—	11.97
1961	3.25	1.08	4.67	1.29	5.96	0.38	NA	0.38	10.66	1.83	NA	NA	—	12.49
1962	3.48	1.21	4.86	1.37	6.23	0.38	NA	0.38	11.30	1.97	NA	NA	—	13.27
1963	3.59	1.27	5.13	1.41	6.55	0.42	NA	0.42	11.83	2.14	NA	NA	—	13.97
1964	3.79	1.37	5.52	1.37	6.89	0.44	NA	0.44	12.49	2.32	NA	NA	—	14.81
1965	3.90	1.44	5.96	1.16	7.11	0.50	NA	0.50	12.96	2.32	NA	NA	—	15.28
1966	4.14	1.62	6.51	1.03	7.55	0.54	NA	0.54	13.84	2.61	NA	NA	—	16.45
1967	4.31	1.96	6.65	1.14	7.79	0.58	NA	0.58	14.64	2.75	NA	NA	—	17.39
1968	4.45	2.08	7.13	1.24	8.37	0.59	NA	0.59	15.48	3.15	NA	NA	—	18.63
1969	4.73	2.25	7.61	1.35	8.96	0.63	NA	0.63	16.57	3.49	NA	NA	—	20.06
1970	4.84	2.40	7.85	1.40	9.25	0.72	NA	0.72	17.21	3.93	NA	NA	—	21.14
1971	4.97	2.51	8.18	1.41	9.59	0.74	NA	0.74	17.82	3.98	NA	NA	—	21.79
1972	5.13	2.61	8.17	1.46	9.62	0.77	NA	0.77	18.12	3.98	NA	NA	—	22.10
1973	4.88	2.60	8.69	1.50	10.18	0.73	NA	0.73	18.39	3.66	NA	NA	—	22.05
1974	4.79	2.56	8.29	1.48	9.77	0.67	NA	0.67	17.78	3.44	NA	NA	—	21.22
1975	4.92	2.51	6.97	1.40	8.36	0.58	NA	0.58	16.38	3.16	NA	NA	—	19.54
1976	5.05	2.67	6.96	1.63	8.60	0.55	NA	0.55	16.87	3.08	NA	NA	—	19.95
1977	4.82	2.50	6.82	1.66	8.47	0.53	NA	0.53	16.33	3.19	NA	NA	—	19.52
1978	4.90	2.60	6.76	1.65	8.40	0.53	NA	0.53	16.44	3.19	NA	NA	—	19.63
1979	4.97	2.79	6.90	1.50	8.40	0.60	NA	0.60	16.75	3.49	NA	NA	—	20.24
1980	4.75	2.61	7.17	1.03	8.20	0.63	NA	0.63	16.20	3.68	NA	NA	—	19.88
1981	4.55	2.52	7.13	0.93	8.06	0.64	NA	0.64	15.76	3.64	NA	NA	—	19.40
1982	4.63	2.61	5.83	1.11	6.94	0.60	NA	0.60	14.78	3.23	NA	NA	—	18.00
1983	4.38	2.43	5.64	0.98	6.62	0.49	NA	0.49	13.92	2.91	NA	NA	—	16.83
1984	4.56	2.52	6.15	1.08	7.23	0.53	NA	0.53	14.84	3.11	NA	NA	—	17.95
1985	4.43	2.43	5.90	0.97	6.87	0.50	NA	0.50	14.24	3.04	NA	NA	—	17.28
1986	4.31	2.32	5.58	0.92	6.50	0.49	NA	0.49	13.62	2.60	NA	NA	—	16.22
1987	4.31	2.43	5.95	1.15	7.10	0.52	NA	0.52	14.37	2.84	NA	NA	—	17.21
1988	4.63	2.67	6.38	1.10	7.48	0.61	NA	0.61	15.39	2.64	NA	NA	—	18.03

TABLE 3.2

Natural gas consumption by sector, 1949–2000 [CONTINUED]

(trillion cubic feet)

Year	Residential Delivered to residences	Commercial[2] Delivered to commercial facilities	Industrial[2] Delivered to industrial facilities	Industrial[2] Lease and plant fuel	Industrial[2] Total	Transportation Pipeline fuel[3]	Transportation Delivered for vehicle fuel use	Transportation Total	End-Use sectors Total	Electric power sector[1] Electric utilities	Electric power sector[1] Nonutility power producers	Electric power sector[1] Total	End-use/ nonutility adjustment[4]	Total consumption
1991	4.56	2.73	7.23	1.13	8.36	0.60	(s)	0.60	16.25	2.79	NA	NA	—	19.04
1992	4.69	2.80	7.53	1.17	8.70	0.59	(s)	0.59	16.78	2.77	NA	NA	—	19.54
1993	4.96	2.86	7.98	1.17	9.15	0.62	(s)	0.63	17.60	2.68	NA	NA	—	20.28
1994	4.85	2.90	8.17	1.12	9.29	0.69	(s)	0.69	17.72	2.99	NA	NA	—	20.71
1995	4.85	3.03	8.58	1.22	9.80	0.70	(s)	0.70	18.38	3.20	NA	NA	—	21.58
1996	5.24	3.16	8.87	1.25	10.12	0.71	(s)	0.71	19.23	2.73	NA	NA	—	21.97
1997	4.98	3.21	8.83	1.20	10.04	0.75	(s)	0.76	18.99	2.97	NA	NA	—	21.96
1998	4.52	3.00	8.69	1.16	9.84	0.64	0.01	0.64	18.00	3.26	NA	NA	—	21.26
1999	R4.73	R3.04	R9.00	R1.08	R10.08	R0.74	0.01	R0.74	18.59	R3.11	2.57	5.68	-2.57	R21.70
2000P	4.92	3.36	9.39	1.26	10.64	0.75	NA	0.75	19.68	3.03	E3.29	E6.32	-3.29	22.71

[1]Data are for natural gas consumed to produce electricity only; exclude natural gas consumed to produce useful thermal output.

[2]Includes natural gas consumed at nonutilities.

[3]Natural gas consumed in the operation of pipelines, primarily in compressors.

[4]Represents the adjustment necessary to avoid double-counting natural gas consumption at nonutilities, which is included under both "End-Use Sectors" and "Electric Power Sector." Due to differences in presentation of data among various EIA publications, data for consumption at nonutility power producers are sometimes shown in the end-use sectors and at other times in the electric power sector. The "End-Use Sectors" data come from EIA's *Natural Gas Monthly* and *Natural Gas Annual*, which include all nonutility consumption in the end-use sectors (mostly industrial, with a small amount of commercial). The "Electric Power Sector" data come from EIA's *Electric Power Monthly* and *Electric Power Annual*, which for 1999 and 2000 include nonutility consumption for electricity generation. Because of this double-counting of nonutility data in 1999 and 2000, an adjustments column is incorporated into the table to remove the overcount and allow the table's components to be added across to reach the total consumption value.

R=Revised. P=Preliminary. E=Estimate. NA=Not available. (s)=Less than 0.005 trillion cubic feet. — =Not applicable.

Note: Natural gas consumption statistics are compiled from surveys of natural gas production, transmission, and distribution companies and electric utility companies. Consumption by sector from these surveys is compiled on a national and individual state basis and then balanced with national and individual state supply data. Included in the data are the following: Commercial Sector-consumption by nonmanufacturing establishments, by municipalities for institutional heating and lighting, and those engaged in agriculture, forestry, and fishing (through 1995); Electric Power Sector-consumption for the generation of electric power; Industrial Sector-consumption by establishments engaged primarily in processing unfinished materials into another form of product (includes mining, petroleum refining, manufacturing and agriculture, forestry and fishing (beginning in 1996) and natural gas industry use for lease and plant fuel); Residential Sector-consumption by private households for space heating, cooking, and other household uses; Transportation Sector-natural gas transmission (pipeline) fuel, and natural gas delivered for use as vehicle fuel. Beginning with 1965, all volumes are shown on a pressure base of 14.73 p.s.i.a. at 60° F. For prior years, the pressure base was 14.65 p.s.i.a. at 60° F. Totals may not equal sum of components due to independent rounding.

SOURCE: "Table 6.5. Natural Gas Consumption by Sector, 1949–2000," in *Annual Energy Review 2000*, U.S. Department of Energy, Energy Information Administration, Washington, DC, 2001.

FIGURE 3.11

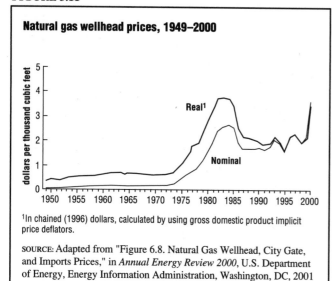

Natural gas wellhead prices, 1949–2000

[1]In chained (1996) dollars, calculated by using gross domestic product implicit price deflators.

SOURCE: Adapted from "Figure 6.8. Natural Gas Wellhead, City Gate, and Imports Prices," in *Annual Energy Review 2000*, U.S. Department of Energy, Energy Information Administration, Washington, DC, 2001

far the largest producers, with Russia accounting for 20.8 trillion cubic feet and the United States producing 18.6 trillion cubic feet.

World Consumption

The world's consumption of natural gas has continued to increase since 1980, from 53 trillion cubic feet to 84 trillion cubic feet in 1999. The United States consumed the largest amount of natural gas in 1999, followed by Russia. (See Figure 3.14.) Combined, they accounted for 42 percent of world consumption.

FUTURE TRENDS IN THE GAS INDUSTRY

The EIA, in *Annual Energy Outlook 2002* (published in 2001), projects energy supply, demand, and prices through 2020. The EIA predicts that natural gas production will increase steadily, as will consumption, pipeline expansion, and imports. Prices, however, are expected to remain steady, declining overall by about 0.5 percent.

Domestic Production

Total domestic natural gas production is projected to increase from 2000 levels of 19.2 trillion cubic feet to 28.6 trillion cubic feet in 2020, with an annual growth rate of about 2 percent. Domestic production will be boosted by increases from onshore sources in the lower 48 states, offshore sources in the Gulf of Mexico, and improvements in exploration and production technology.

Figure 3.15 shows projected increases in different types of natural gas production. The letters "NA" and "AD" in Figure 3.15 stand for nonassociated (NA) and associated-dissolved (AD) natural gas. These terms refer to natural gas that is found in conjunction with crude oil (associated) or not in conjunction with crude oil (nonassociated). Associated-dissolved natural gas is found in a dissolved state with the oil, like oxygen dissolved in aquarium water.

Domestic Consumption

Total domestic natural gas consumption is projected to increase from 2000 levels of 22.8 trillion cubic feet to 33.8 trillion cubic feet in 2020, with an annual growth rate of about 2 percent. Demand for natural gas by electricity producers is expected to account for 55 percent of the total consumption growth by 2020. (See Figure 3.16.) Natural gas will be called upon to replace the nation's aging nuclear electricity plants because of its efficiency and low emissions. To meet growing demand, natural gas pipeline capacity will have to be expanded, particularly along the corridors that move Canadian supplies to the Pacific Coast, and along corridors that move Pacific Coast Canadian imports and Gulf offshore natural gas to the East.

Imports and Exports

Net imports of natural gas are projected to increase to meet demand, from 3.5 trillion cubic feet in 2000 to 5.5 trillion cubic feet in 2020. Most of these imports will come from Canada. (See Figure 3.17.) The rest will be shipped to the United States from Algeria, Australia, Indonesia, Nigeria, Oman, Qatar, Trinidad, and the United Arab Emirates in the form of liquefied natural gas. Natural gas exports to Mexico are expected to continue and to peak in 2015. After that time exports to Mexico are expected to decline; Mexico's own natural gas infrastructure should be developed by 2015, and it is expected to begin to meet its own natural gas needs.

Prices

The EIA predicts that natural gas prices will remain relatively steady from 2000 to 2020 in all sectors: industrial, electricity, transportation, residential, and commercial. In fact, prices could fall overall by about 0.5 percent. The average wellhead gas price is also projected to fall by about 0.5 percent from 2000 to 2020.

TABLE 3.3

Natural gas prices by sector, 1967–2000

(Price: dollars per thousand cubic feet; share of total volume delivered: percentage)

Year	Residential Prices[4] Nominal	Real[6]	Commercial[1] Prices Nominal	Real[6]	Share of total volume delivered	Industrial[2] Prices Nominal	Real[6]	Share of total volume delivered	Vehicle fuel[3] Prices Nominal	Real[6]	Share of total volume delivered	Electric utilities Prices[5] Nominal	Real[6]
1967	1.04	4.13	0.74	2.94	NA	0.34	1.35	NA	NA	NA	NA	0.28	1.11
1968	1.04	3.95	0.73	2.78	NA	0.34	1.29	NA	NA	NA	NA	0.22	0.84
1969	1.05	3.81	0.74	2.68	NA	0.35	1.27	NA	NA	NA	NA	0.27	0.98
1970	1.09	3.75	0.77	2.65	NA	0.37	1.27	NA	NA	NA	NA	0.29	1.00
1971	1.15	3.77	0.82	2.69	NA	0.41	1.34	NA	NA	NA	NA	0.32	1.05
1972	1.21	3.80	0.88	2.77	NA	0.45	1.41	NA	NA	NA	NA	0.34	1.07
1973	1.29	3.84	0.94	2.80	NA	0.50	1.49	NA	NA	NA	NA	0.38	1.13
1974	1.43	3.90	1.07	2.92	NA	0.67	1.83	NA	NA	NA	NA	0.51	1.39
1975	1.71	4.27	1.35	3.37	NA	0.96	2.40	NA	NA	NA	NA	0.77	1.92
1976	1.98	4.68	1.64	3.88	NA	1.24	2.93	NA	NA	NA	NA	1.06	2.51
1977	2.35	5.22	2.04	4.53	NA	1.50	3.33	NA	NA	NA	NA	1.32	2.93
1978	2.56	5.31	2.23	4.62	NA	1.70	3.52	NA	NA	NA	NA	1.48	3.07
1979	2.98	5.70	2.73	5.22	NA	1.99	3.81	NA	NA	NA	NA	1.81	3.46
1980	3.68	6.45	3.39	5.94	NA	2.56	4.49	NA	NA	NA	NA	2.27	3.98
1981	4.29	6.88	4.00	6.41	NA	3.14	5.03	NA	NA	NA	NA	2.89	4.63
1982	5.17	7.80	4.82	7.28	NA	3.87	5.84	85.1	NA	NA	NA	3.48	5.25
1983	6.06	8.80	5.59	8.12	NA	4.18	6.07	80.7	NA	NA	NA	3.58	5.20
1984	6.12	8.57	5.55	7.77	NA	4.22	5.91	74.7	NA	NA	NA	3.70	5.18
1985	6.12	8.31	5.50	7.46	NA	3.95	5.36	68.8	NA	NA	NA	3.55	4.82
1986	5.83	7.74	5.08	6.75	NA	3.23	4.29	59.8	NA	NA	NA	2.43	3.23
1987	5.54	7.14	4.77	6.15	93.1	2.94	3.79	47.4	NA	NA	NA	2.32	2.99
1988	5.47	6.82	4.63	5.77	90.7	2.95	3.68	42.6	NA	NA	NA	2.33	2.90
1989	5.64	6.77	4.74	5.69	89.1	2.96	3.55	36.9	NA	NA	NA	2.43	2.92
1990	5.80	6.70	4.83	5.58	86.6	2.93	3.39	35.2	3.39	3.92	NA	2.38	2.75
1991	5.82	6.49	4.81	5.36	85.1	2.69	3.00	32.7	3.96	4.42	NA	2.18	2.43
1992	5.89	6.41	4.88	5.31	83.2	2.84	3.09	30.3	4.05	4.41	NA	2.36	2.57
1993	6.16	6.55	5.22	5.55	83.9	3.07	3.26	29.7	4.27	4.54	87.8	2.61	2.78
1994	6.41	6.68	5.44	5.67	79.3	3.05	3.18	25.5	4.11	4.28	86.9	2.28	2.37
1995	6.06	6.18	5.05	5.15	76.7	2.71	2.76	24.5	3.98	4.06	86.6	2.02	2.06
1996	6.34	6.34	5.40	5.40	77.6	3.42	3.42	19.4	4.34	4.34	94.0	2.69	2.69
1997	6.94	6.81	5.80	5.69	70.8	3.59	3.52	18.1	4.44	4.36	89.7	2.78	2.73
1998	6.82	6.61	5.48	5.31	67.0	3.14	R3.04	16.1	4.59	4.45	85.4	2.40	2.33
1999	R6.69	R6.39	R5.33	R5.09	R66.2	R3.10	R2.96	R17.4	R4.34	R4.14	R85.6	R2.62	R2.50
2000P	7.70	7.20	6.16	5.76	64.2	4.45	4.16	15.6	NA	NA	NA	NA	NA

[1]Includes deliveries to municipalities and public authorities for institutional heating and other purposes.
[2]Most volumes and associated revenues for deliveries to nonutility power producers are included in the industrial sector. In instances where the nonutility is primarily a commercial establishment, volumes and associated revenues are included in the calculation of commercial prices.
[3]Much of the natural gas delivered for vehicle fuel represents deliveries to fueling stations that are used primarily or exclusively by respondents' fleet vehicles. Thus, the prices are often those associated with the operation of fleet vehicles.
[4]Based on 100 percent of volume delivered.
[5]Based on all steam-electric utility plants with a combined capacity of 50 megawatts or greater.
[6]In chained (1996) dollars, calculated by using gross domestic product implicit price deflators.
R=Revised. P=Preliminary. NA=Not available.
Notes: Natural gas includes supplemental gaseous fuels. Residential, commercial, and industrial price data represent prices of natural gas sold and delivered by local distribution companies to residential, commercial, and industrial consumers, respectively. The data do not reflect prices of natural gas transported for the account of others. The average for each end-use sector is calculated by dividing the total value of the gas consumed by each sector by the total quantity consumed.

SOURCE: "Table 6.9 Natural Gas Prices by Sector, 1967–2000 (Price: Dollars per Thousand Cubic Feet; Share of Total Volume Delivered: Percentage)," in *Annual Energy Review 2000*, U.S. Department of Energy, Energy Information Administration, Washington, DC, 2001

FIGURE 3.12

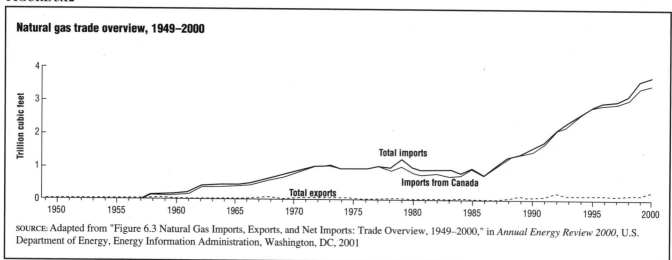

Natural gas trade overview, 1949–2000

SOURCE: Adapted from "Figure 6.3 Natural Gas Imports, Exports, and Net Imports: Trade Overview, 1949–2000," in *Annual Energy Review 2000*, U.S. Department of Energy, Energy Information Administration, Washington, DC, 2001

FIGURE 3.13

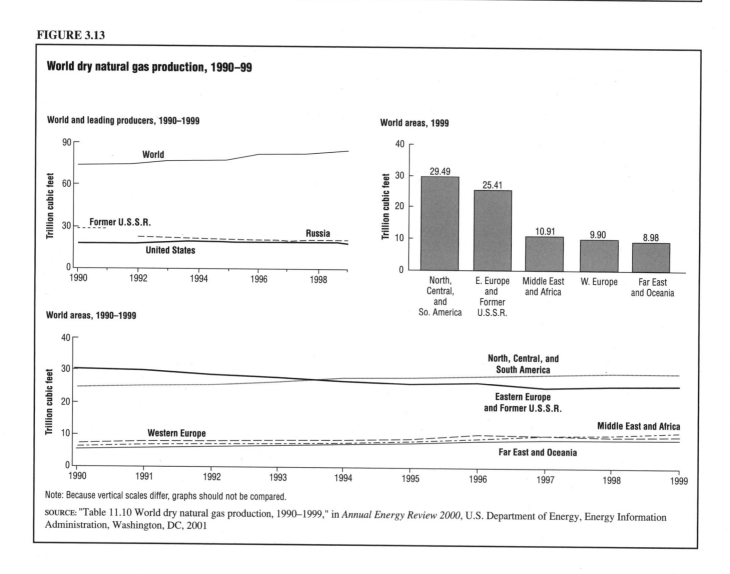

World dry natural gas production, 1990–99

Note: Because vertical scales differ, graphs should not be compared.

SOURCE: "Table 11.10 World dry natural gas production, 1990–1999," in *Annual Energy Review 2000*, U.S. Department of Energy, Energy Information Administration, Washington, DC, 2001

FIGURE 3.14

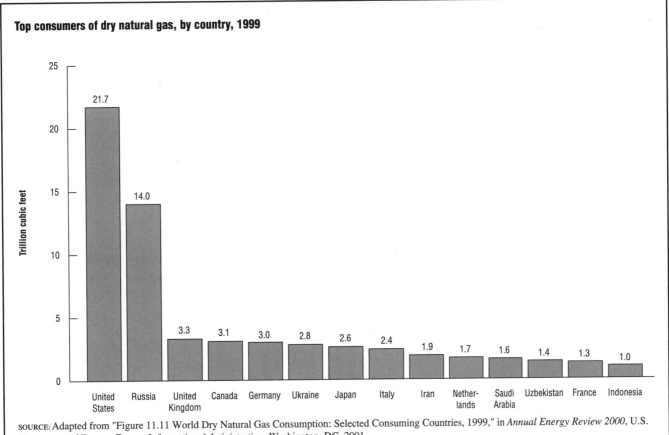

Top consumers of dry natural gas, by country, 1999

SOURCE: Adapted from "Figure 11.11 World Dry Natural Gas Consumption: Selected Consuming Countries, 1999," in *Annual Energy Review 2000*, U.S. Department of Energy, Energy Information Administration, Washington, DC, 2001

FIGURE 3.15

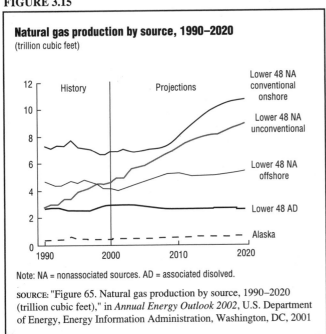

Natural gas production by source, 1990–2020
(trillion cubic feet)

Note: NA = nonassociated sources. AD = associated disolved.

SOURCE: "Figure 65. Natural gas production by source, 1990–2020 (trillion cubic feet)," in *Annual Energy Outlook 2002*, U.S. Department of Energy, Energy Information Administration, Washington, DC, 2001

FIGURE 3.16

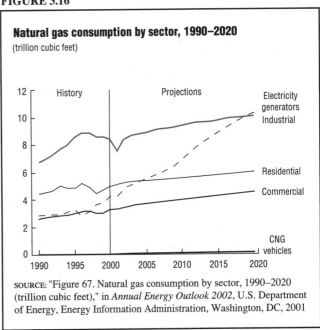

Natural gas consumption by sector, 1990–2020
(trillion cubic feet)

SOURCE: "Figure 67. Natural gas consumption by sector, 1990–2020 (trillion cubic feet)," in *Annual Energy Outlook 2002*, U.S. Department of Energy, Energy Information Administration, Washington, DC, 2001

FIGURE 3.17

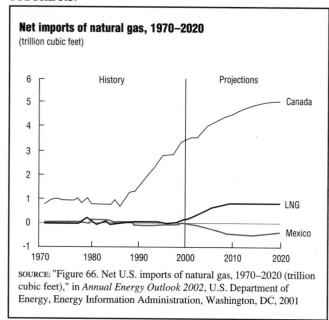

Net imports of natural gas, 1970–2020
(trillion cubic feet)

SOURCE: "Figure 66. Net U.S. imports of natural gas, 1970–2020 (trillion cubic feet)," in *Annual Energy Outlook 2002*, U.S. Department of Energy, Energy Information Administration, Washington, DC, 2001

CHAPTER 4

COAL

A HISTORICAL PERSPECTIVE

The first large-scale use of coal occurred during the Industrial Revolution in England. At that time the sky was filled with billowing columns of black smoke, soot covered the towns and cities, and workers breathed the thick coal dust swirling around them. Most people then were not concerned with environmental issues because the Industrial Revolution meant jobs to the workers, and factory owners had little desire to control the pollution their factories were creating. In addition, environmental and public health considerations were not as well understood as they are today.

In America early colonists used wood to heat their homes because it was so plentiful and coal was less available. Prior to the Civil War (1861–65), some industries used coal as a source of energy, but its major use began with the building of the railroads. After the Civil War ended, America began to expand its railway system westward and to increase its manufacturing capacity. Coal became such a fundamental part of American industrialization that some have called this era "the Coal Age." As in England, Americans considered the development of industry a source of national pride. Photographs and postcards of the time proudly featured railroad trains and steel mills with smokestacks belching dark smoke into gray skies.

By the 1900s coal had become America's major fuel source, accounting for nearly 90 percent of the nation's energy requirements. By the end of World War II (1939–45), however, coal accounted for only 38 percent of the energy supply, as oil began to heat homes and offices, and the growing number of cars used gasoline. Coal fell further out of favor as an energy source in the 1950s and 1960s as oil became more attractive as a cleaner fuel for heating homes and businesses. The decline of coal use continued, with coal producing as little as 18 percent of the energy used during some years in the early 1970s because of concerns about environmental pollution and the emergence of nuclear power as an energy source.

By 1973, however, Americans recognized they could no longer rely on imported oil for their energy. The Arab oil embargo clearly demonstrated the nation's heavy reliance on foreign sources of energy and its potentially crippling effect on the American economy. Consequently, the nation revived its interest in domestic coal as a plentiful and economical energy source.

After the 1973 embargo, coal and nuclear fuel received more attention, especially in the electric utility sector. In 1977 President Jimmy Carter called for a two-thirds annual increase in national coal production by 1985. He also asked utility companies and other large industries to convert their operations to coal and proposed a 10-year, $10 billion program to encourage domestic coal production. In 2000 coal contributed more to America's energy production than any other source, generating 23 quadrillion Btu and 32 percent of all energy. (See Figure 4.1.)

WHAT IS COAL?

Coal is a black, combustible, mineral solid that develops over a period of millions of years from the partial decomposition of plant matter in an airless space, under increased temperature and pressure. Coal beds, sometimes called seams, are found in the earth between beds of sandstone, shale, and limestone and range in thickness from less than an inch to more than one hundred feet. Approximately five to ten feet of ancient, composted plant material have been compressed to create each foot of coal.

Coal is used as a fuel and in the production of coke (the solid substance left after coal gas and coal tar have been extracted), coal gas, water gas, and many coal-tar compounds. When coal is burned, its fossil energy—sunlight converted and stored by plants over millions of years—is released. One ton of coal produces an average of 22 million Btu, about the same heating value as 22,000 cubic feet of natural gas, 159 gallons of distillate fuel oil,

FIGURE 4.1

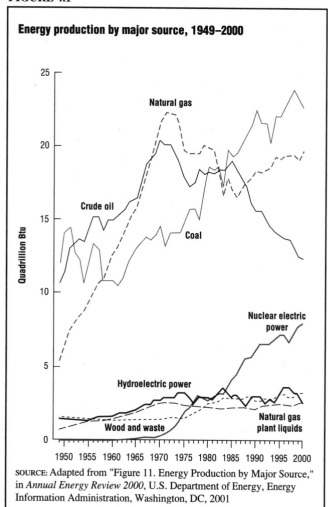

Energy production by major source, 1949–2000

SOURCE: Adapted from "Figure 11. Energy Production by Major Source," in *Annual Energy Review 2000*, U.S. Department of Energy, Energy Information Administration, Washington, DC, 2001

or one cord of seasoned firewood. (A cord is a stack of wood 4 feet by 4 feet by 8 feet, or 128 cubic feet.)

CLASSIFICATIONS OF COAL

There are four basic types of coal. Classifications, or "coal ranks," are based on how much carbon, volatile matter, and heating value are contained in the coal.

- Anthracite, or hard coal, is the highest ranked coal. It is hard and jet black, with a moisture content of less than 15 percent. Anthracite is used mainly for generating electricity and for space heating. It contains approximately 22–28 million Btu per ton, with an ignition temperature of approximately 925–970 degrees Fahrenheit. Anthracite is mined mainly in northeastern Pennsylvania. (See Figure 4.2.)

- Bituminous, or soft coal, is the most common coal. It is dense and black, with a moisture content of less than 20 percent and an ignition range of 700–900 degrees Fahrenheit. Bituminous coal is used to generate electricity, for space heating, and to produce coke. Bituminous coal contains a heating value range of

19–30 million Btu per ton. It is mined chiefly in the Appalachian and Midwest regions of the United States. (See Figure 4.2.)

- Subbituminous coal, or black lignite, is dull black in color and generally contains 20 to 30 percent moisture. Black lignite is used for generating electricity and for space heating. It contains 16–24 million Btu per ton. Black lignite is mined primarily in the western United States. (See Figure 4.2.)

- Lignite, the lowest ranked coal, is brownish-black in color and has a high moisture content. It tends to disintegrate when exposed to weather. Lignite is used mainly to generate electricity and contains about 9–17 million Btu per ton. Lignite has an ignition temperature of approximately 600 degrees Fahrenheit. Most lignite is mined in North Dakota, Montana, Texas, California, and Louisiana. (See Figure 4.2.)

Bituminous coal accounts for, by far, the largest share of all coal production; subbituminous is second. (See Table 4.1.) In 2000 production of all types of coal totaled nearly 1.1 billion short tons. (A short ton of coal is 2,000 pounds.) Of that, nearly 1 billion short tons were bituminous and subbituminous coal. Lignite and anthracite accounted for the remainder.

LOCATIONS OF COAL DEPOSITS

Coal is found in about 13 percent, or 458,600 square miles, of the U.S. total land area. (See Figure 4.2.) Geologists have geographically divided U.S. coalfields into three zones: the Appalachian, Interior, and Western regions. The Appalachian region is subdivided into three areas: Northern (Ohio, Pennsylvania, Maryland, and northern West Virginia); Central (Virginia, southern West Virginia, eastern Kentucky, and Tennessee); and Southern Appalachia (Alabama). Coal production in the Interior region occurs in Illinois, Indiana, western Kentucky, Iowa, Missouri, Kansas, Arkansas, Oklahoma, Louisiana, and Texas. The Western region includes the Northern Great Plains (Montana, Wyoming, northern Colorado, and North and South Dakota), the Rocky Mountains, the Southwest (southern Colorado, Utah, Arizona, and New Mexico), and the Northwest (Washington and Alaska).

Historically, more coal has been mined east of the Mississippi than west of the Mississippi, but the West's proportion of total production has increased almost every year since 1965. (See Table 4.1.) In 1965 the production of coal in the West was 27 million short tons, only 5 percent of total production. By 1999 western production had increased over 20-fold, to 571 million short tons, or 52 percent of the total. The amount of coal mined east of the Mississippi that year was 530 million short tons. In 2000 slightly over 566 million short tons of coal were mined west of the Mississippi, while slightly over 509 million

FIGURE 4.2

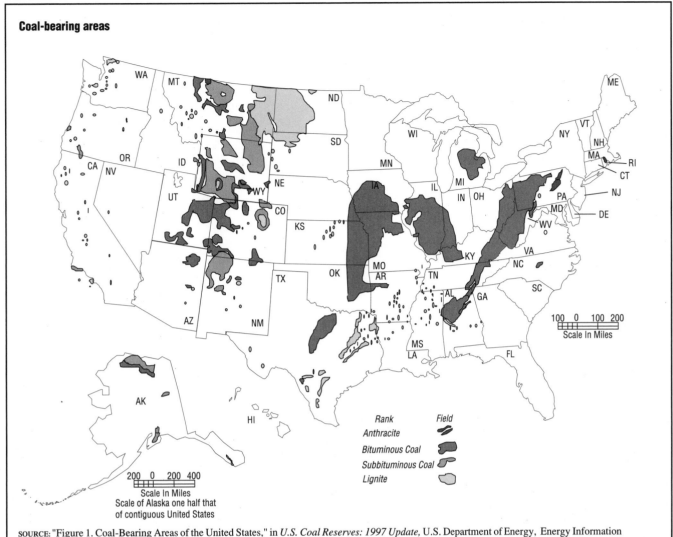

Coal-bearing areas

SOURCE: "Figure 1. Coal-Bearing Areas of the United States," in *U.S. Coal Reserves: 1997 Update,* U.S. Department of Energy, Energy Information Administration, Washington, DC, 1999

short tons were mined to the east. Western production neared 53 percent of the total mined.

The growth in coal production in the West has been partly the result of environmental concerns that have led to an increased demand for low-sulfur coal, which is concentrated in the West. In addition, surface mining, which is cheaper and more efficient, is more prevalent in the West. Finally, improved rail service has made it easier to deliver this low-sulfur coal to utility plants located east of the Mississippi River.

COAL MINING METHODS

The method used to mine coal depends on the terrain and the depth of the coal. Prior to the early 1970s, most coal was taken from underground mines. Since that time, however, coal production has shifted from underground mines to surface mines. (See Table 4.1 and Figure 4.3.)

Underground mining is required when the coal lies deeper than 200 feet below ground level. The depth of

most underground mines is less than 1,000 feet, but a few go down as far as 2,000 feet. In underground mines, some coal must be left untouched in order to form pillars that prevent the mine from caving in. In both underground mines and surface mines, natural features such as folded, faulted, and interlaid rock strata reduce the amount of coal that can be recovered.

Surface mines are usually less than 200 feet deep and can be developed in flat or hilly terrain. Area surface mining is practiced on large plots of relatively flat ground, while contour surface mining follows coal beds along hillsides. (See Figure 4.4.) Open pit mining is used to mine thick, steeply inclined coal beds and uses a combination of contour and area mining methods.

The growing prevalence of surface coal mining and the closing of nonproductive mines led to increases in coal mining productivity through the 1980s and 1990s. (See Figure 4.5.) In 1999 average productivity reached an all-time high of 6.5 short tons per miner hour. Because

TABLE 4.1

Coal production, 1949–2000

(million short tons)

Year	Rank				Mining Method		Location		Total
	Bituminous Coal	Subbituminous Coal	Lignite	Anthracite	Underground	Surface	West of the Mississippi	East of the Mississippi	
1949	437.9	(1)	(1)	42.7	358.9	121.7	36.4	444.2	480.6
1950	516.3	(1)	(1)	44.1	421.0	139.4	36.0	524.4	560.4
1951	533.7	(1)	(1)	42.7	442.2	134.2	34.6	541.7	576.3
1952	466.8	(1)	(1)	40.6	381.2	126.3	32.7	474.8	507.4
1953	457.3	(1)	(1)	30.9	367.4	120.8	30.6	457.7	488.2
1954	391.7	(1)	(1)	29.1	306.0	114.8	25.4	395.4	420.8
1955	464.6	(1)	(1)	26.2	358.0	132.9	26.6	464.2	490.8
1956	500.9	(1)	(1)	28.9	380.8	148.9	25.8	504.0	529.8
1957	492.7	(1)	(1)	25.3	373.6	144.5	24.7	493.4	518.0
1958	410.4	(1)	(1)	21.2	297.6	134.0	20.3	411.3	431.6
1959	412.0	(1)	(1)	20.6	292.8	139.8	20.3	412.4	432.7
1960	415.5	(1)	(1)	18.8	292.6	141.7	21.3	413.0	434.3
1961	403.0	(1)	(1)	17.4	279.6	140.9	21.8	398.6	420.4
1962	422.1	(1)	(1)	16.9	287.9	151.1	21.4	417.6	439.0
1963	458.9	(1)	(1)	18.3	309.0	168.2	23.7	453.5	477.2
1964	487.0	(1)	(1)	17.2	327.7	176.5	25.7	478.5	504.2
1965	512.1	(1)	(1)	14.9	338.0	189.0	27.4	499.5	527.0
1966	533.9	(1)	(1)	12.9	342.6	204.2	28.0	518.8	546.8
1967	552.6	(1)	(1)	12.3	352.4	212.5	28.9	536.0	564.9
1968	545.2	(1)	(1)	11.5	346.6	210.1	29.7	527.0	556.7
1969	547.2	8.3	5.0	10.5	349.2	221.7	33.3	537.7	571.0
1970	578.5	16.4	8.0	9.7	340.5	272.1	44.9	567.8	612.7
1971	521.3	22.2	8.7	8.7	277.2	283.7	51.0	509.9	560.9
1972	556.8	27.5	11.0	7.1	305.0	297.4	64.3	538.2	602.5
1973	543.5	33.9	14.3	6.8	300.1	298.5	76.4	522.1	598.6
1974	545.7	42.2	15.5	6.6	278.0	332.1	91.9	518.1	610.0
1975	577.5	51.1	19.8	6.2	293.5	361.2	110.9	543.7	654.6
1976	588.4	64.8	25.5	6.2	295.5	389.4	136.1	548.8	684.9
1977	581.0	82.1	28.2	5.9	266.6	430.6	163.9	533.3	697.2
1978	534.0	96.8	34.4	5.0	242.8	427.4	183.0	487.2	670.2
1979	612.3	121.5	42.5	4.8	320.9	460.2	221.4	559.7	781.1
1980	628.8	147.7	47.2	6.1	337.5	492.2	251.0	578.7	829.7
1981	608.0	159.7	50.7	5.4	316.5	507.3	269.9	553.9	823.8
1982	620.2	160.9	52.4	4.6	339.2	499.0	273.9	564.3	838.1
1983	568.6	151.0	58.3	4.1	300.4	481.7	274.7	507.4	782.1
1984	649.5	179.2	63.1	4.2	352.1	543.9	308.3	587.6	895.9
1985	613.9	192.7	72.4	4.7	350.8	532.8	324.9	558.7	883.6
1986	620.1	189.6	76.4	4.3	360.4	529.9	325.9	564.4	890.3
1987	636.6	200.2	78.4	3.6	372.9	545.9	336.8	581.9	918.8
1988	638.1	223.5	85.1	3.6	382.2	568.1	370.7	579.6	950.3
1989	659.8	231.2	86.4	3.3	393.8	586.9	381.7	599.0	980.7
1990	693.2	244.3	88.1	3.5	424.5	604.5	398.9	630.2	1,029.1
1991	650.7	255.3	86.5	3.4	407.2	588.8	404.7	591.3	996.0
1992	651.8	252.2	90.1	3.5	407.2	590.3	409.0	588.6	997.5
1993	576.7	274.9	89.5	4.3	351.1	594.4	429.2	516.2	945.4
1994	640.3	300.5	88.1	4.6	399.1	634.4	467.2	566.3	1,033.5
1995	613.8	328.0	86.5	4.7	396.2	636.7	488.7	544.2	1,033.0
1996	630.7	340.3	88.1	4.8	409.8	654.0	500.2	563.7	1,063.9
1997	653.8	345.1	86.3	4.7	420.7	669.3	510.6	579.4	1,089.9
1998	R640.4	R386.1	85.8	5.3	417.7	699.8	547.0	570.6	1,117.5
1999	R596.3	R412.2	R87.2	R4.8	R391.8	R708.6	R570.9	R529.6	R1,100.4
2000	E548.5	E433.8	E88.7	E4.5	E382.9	E692.6	E566.2	E509.3	P1,075.5

[1]Included in bituminous coal.
R=Revised. P=Preliminary. E=Estimate.
Note: Totals may not equal sum of components due to independent rounding.

SOURCE: "Table 7.2 Coal Production, 1949–2000 (Million Short Tons)," in *Annual Energy Review 2000*, U.S. Department of Energy, Energy Information Administration, Washington, DC, 2001

surface mines are easier to work, they average up to three times the productivity of underground mines. In 1999 the productivity for surface mines was 10.3 short tons of coal per miner hour, while underground mines produced 3.9 short tons per miner hour.

COAL IN THE DOMESTIC MARKET

Overall Production and Consumption

Before 1951 coal was the leading source of energy produced in the United States. (See Figure 4.1.) In the years from 1952 to 1984, crude oil and natural gas exchanged first place twice in energy production. After the 1973 oil embargo, however, the Carter administration called for a two-thirds annual increase in national coal production by 1985 and asked utility companies and other large industries to convert their operations to coal. By 1984 coal had regained the top position as the leading source of energy produced in the United States and has remained the leader ever since.

The Energy Information Administration (EIA) of the U.S. Department of Energy (DOE), in its *Annual Energy Review 2000* (published in 2001), notes that the nation consumed 558 million short tons of coal in 1974. Twenty-six years later, in 2000, consumption had grown to nearly 1.1 billion short tons. (Figure 4.6 shows the flow of coal in 2000.) The increases in coal consumption were greatest in the electric utility sector, as many existing electric power plants switched to coal from more expensive oil and gas, and many new, coal-fired power plants were constructed during the 1970s.

Coal Consumption by Sector

To make electricity, coal is pulverized and burned to produce steam, which then drives electric generators.

Each ton of coal used by an electric generator produces about 2,000 kilowatt-hours of electricity. In household terms, each pound of coal produces enough electricity to light ten 100-watt lightbulbs for one hour.

Electric utility companies are by far the largest consumers of coal today. (See Figure 4.6 and Figure 4.7.)

FIGURE 4.3

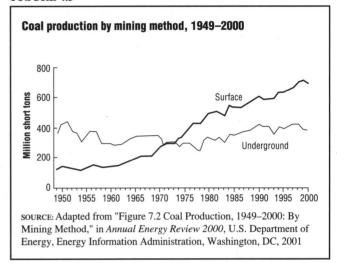

Coal production by mining method, 1949–2000

SOURCE: Adapted from "Figure 7.2 Coal Production, 1949–2000: By Mining Method," in *Annual Energy Review 2000*, U.S. Department of Energy, Energy Information Administration, Washington, DC, 2001

FIGURE 4.4

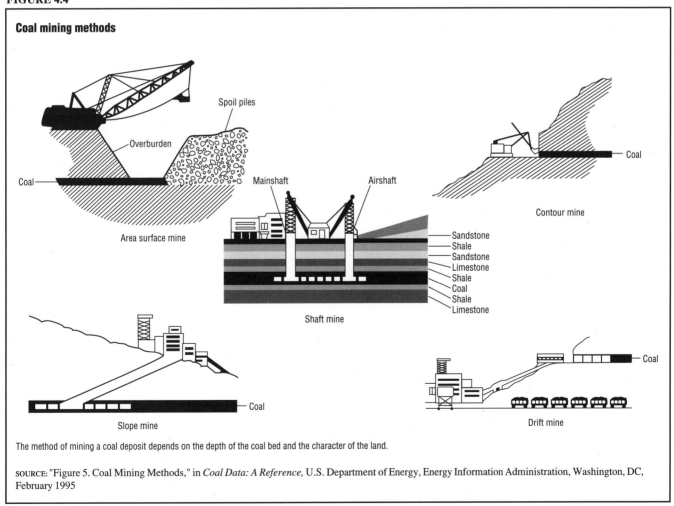

Coal mining methods

The method of mining a coal deposit depends on the depth of the coal bed and the character of the land.

SOURCE: "Figure 5. Coal Mining Methods," in *Coal Data: A Reference,* U.S. Department of Energy, Energy Information Administration, Washington, DC, February 1995

They accounted for 92 percent of domestic coal consumption, or 991 million short tons, in 2000. Coal-fired plants produced nearly 2 trillion kilowatt-hours of electricity, or 52 percent of U.S. electricity net generation, in 2000.

The industrial sector was the second-largest consumer of coal in 2000, accounting for 8 percent of coal use, or 87.8 million short tons. Coal is used in many industrial applications, including the chemical, cement, paper, synthetic fuels, metals, and food-processing industries.

Coal was once the major fuel in the residential and commercial sector. (See Figure 4.7.) After the late 1940s, however, coal was replaced by oil, natural gas, and electricity, which are cleaner and more convenient. By 1970 only 16 million tons of coal were used for residential and commercial buildings. Since then, residential and commercial coal use has continued to decline, falling to 0.6 million short tons in 2000, or far less than 1 percent of total coal use.

The Price of Coal

In 1999 the average price of coal fell to $16 per short ton, down for the 20th year in a row and only 35 percent of the 1979 price in real dollars, which are adjusted for inflation. (See Table 4.2.) On a per-Btu basis, coal remains the least expensive fossil fuel. In 1999 the average cost of coal was 80 cents per million Btu, compared with $1.86 per million Btu for natural gas and $2.56 per million Btu for crude oil.

ENVIRONMENTAL CONCERNS ABOUT COAL

Problems

The negative side of energy use—pollution of the environment—is not a recent problem. In 1306 King Edward I of England so objected to the noxious smoke from London's coal-burning fires that he banned coal's use by everyone except blacksmiths. The enormous

FIGURE 4.5

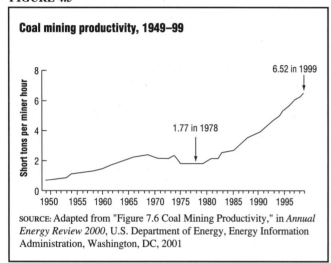

SOURCE: Adapted from "Figure 7.6 Coal Mining Productivity," in *Annual Energy Review 2000*, U.S. Department of Energy, Energy Information Administration, Washington, DC, 2001

FIGURE 4.6

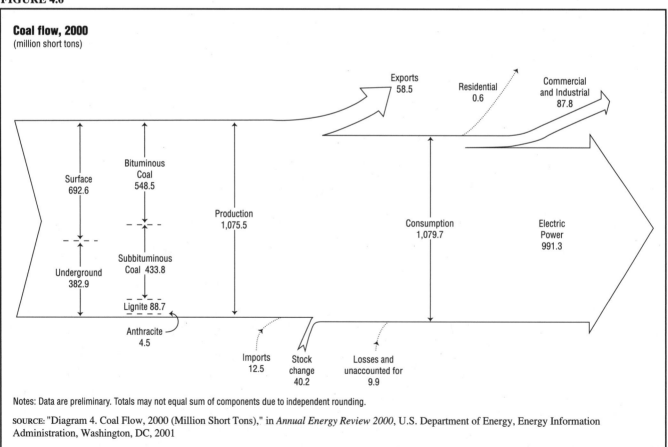

Notes: Data are preliminary. Totals may not equal sum of components due to independent rounding.

SOURCE: "Diagram 4. Coal Flow, 2000 (Million Short Tons)," in *Annual Energy Review 2000*, U.S. Department of Energy, Energy Information Administration, Washington, DC, 2001

FIGURE 4.7

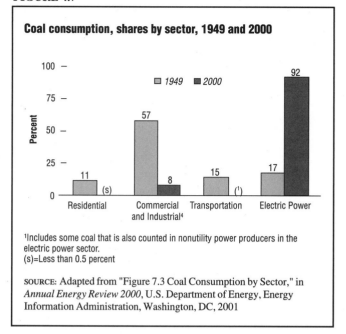

Coal consumption, shares by sector, 1949 and 2000

[1]Includes some coal that is also counted in nonutility power producers in the electric power sector.
(s)=Less than 0.5 percent

SOURCE: Adapted from "Figure 7.3 Coal Consumption by Sector," in *Annual Energy Review 2000*, U.S. Department of Energy, Energy Information Administration, Washington, DC, 2001

FIGURE 4.8

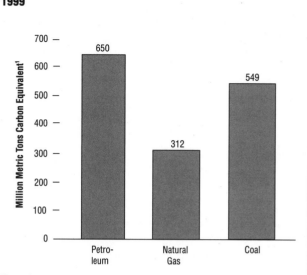

Carbon dioxide emissions from energy consumption, by fuel, 1999

[1]Tons of carbon equivalent can be converted to tons of carbon dioxide gas by multiplying by 3.667. One ton of carbon equivalent = 3.667 tons of carbon dioxide gas.

SOURCE: Adapted from "Figure 12.3 Carbon Dioxide Emissions From Energy Consumption by Sector by Energy Source, 1999: by Fuel," in *Annual Energy Review 2000*, U.S. Department of Energy, Energy Information Administration, Washington, DC, 2001

scale of today's energy use has increased environmental concerns.

Coal-fired electric power plants emit gases that are harmful to the environment. Scientists believe that burning huge quantities of fossil fuels causes the "greenhouse effect," in which "greenhouse" gases from fuels trap heat in the earth's atmosphere and cause increased warming, which may threaten the environment. Burning coal also contributes to the formation of acid rain and to public health concerns. Sulfur dioxide, for instance, has been shown to cause respiratory problems.

Carbon dioxide accounts for the largest share of greenhouse gas emissions. In 1999 the combustion of coal in the United States produced more than a half billion metric tons of carbon equivalent, or 36 percent of total carbon dioxide emissions from all sources. (See Figure 4.8.)

ACID RAIN. Acid rain is any form of precipitation that contains a greater-than-normal amount of acid. Even non-polluted rain is slightly acidic (with a pH of about 5.6) because rainwater combines with the carbon dioxide normally found in the air to produce a weak acid called carbonic acid. However, pollutants in the air can increase the acidity of rain and other forms of precipitation, such as snow and fog.

Chemicals such as oxides of sulfur and nitrogen, which are given off during the combustion of fossil fuels, are pollutants that combine with precipitation to form acids. These oxides increase in the air because of automobile exhaust, industrial and power plant emissions, and other fossil fuel combustion processes. In many parts of the world acid rain has caused significant damage to forests, lakes, and other ecosystems.

Solutions

THE CLEAN COAL TECHNOLOGY LAW. In 1984 Congress established the DOE's Clean Coal Technology (CCT) program (PL 98-473). Congress directed the DOE to administer cost-shared projects (financed by both industry and government) to demonstrate clean coal technologies. The demonstration projects had the goal of using coal in more environmentally and economically efficient ways.

CLEAN COAL TECHNOLOGY AND THE CLEAN AIR ACT. The stated goal of both Congress and the DOE has been to develop cost-effective ways to burn coal more cleanly, both to control acid rain and to improve the nation's energy security by reducing dependence on imported fuels. One strategy is a slow, phased-in approach in which utility companies and states reduce their emissions in stages.

Under the Clean Air Act of 1990 (PL 101-549), restrictions on sulfur dioxide and nitrogen oxide emissions took effect in 1995 and tightened in 2000. Each round of regulation requires coal-burning utilities to find lower-sulfur coal or to install cleaner technology, such as "scrubbers" that reduce smokestack emissions, which contribute heavily to air pollution. When the first Clean Air Act was passed in 1970, it was aimed at changing the air-quality standards at new generating stations, and older coal-using plants were exempt. Under the 1990 act, older plants are also covered by the regulations.

TABLE 4.2

Coal prices, 1949–1999
(dollars per short ton)

Year	Bituminous Coal Nominal	Bituminous Coal Real[2]	Subbituminous Coal Nominal	Subbituminous Coal Real[2]	Lignite Nominal	Lignite Real[2]	Subtotal[1] Nominal	Subtotal[1] Real[2]	Anthracite Nominal	Anthracite Real[2]	Total Nominal	Total Real[2]
1949	[3]4.90	[3]28.39	(4)	(4)	2.37	13.73	4.88	28.27	8.90	51.56	5.24	30.36
1950	[3]4.86	[3]27.85	(4)	(4)	2.41	13.81	4.84	27.74	9.34	53.52	5.19	29.74
1951	[3]4.94	[3]26.40	(4)	(4)	2.44	13.04	4.92	26.30	9.94	53.13	5.29	28.27
1952	[3]4.92	[3]25.89	(4)	(4)	2.39	12.58	4.90	25.79	9.58	50.42	5.27	27.74
1953	[3]4.94	[3]25.66	(4)	(4)	2.38	12.36	4.92	25.56	9.87	51.27	5.23	27.17
1954	[3]4.54	[3]23.35	(4)	(4)	2.43	12.50	4.52	23.25	8.76	45.06	4.81	24.74
1955	[3]4.51	[3]22.80	(4)	(4)	2.38	12.03	4.50	22.75	8.00	40.44	4.69	23.71
1956	[3]4.83	[3]23.62	(4)	(4)	2.39	11.69	4.82	23.57	8.33	40.73	5.01	24.50
1957	[3]5.09	[3]24.09	(4)	(4)	2.35	11.12	5.08	24.04	9.11	43.11	5.28	24.99
1958	[3]4.87	[3]22.50	(4)	(4)	2.35	10.86	4.86	22.46	9.14	42.24	5.07	23.43
1959	[3]4.79	[3]21.89	(4)	(4)	2.25	10.28	4.77	21.80	8.55	39.08	4.95	22.62
1960	[3]4.71	[3]21.23	(4)	(4)	2.29	10.32	4.69	21.14	8.01	36.10	4.83	21.77
1961	[3]4.60	[3]20.50	(4)	(4)	2.24	9.98	4.58	20.41	8.26	36.81	4.73	21.08
1962	[3]4.50	[3]19.79	(4)	(4)	2.23	9.81	4.48	19.70	7.99	35.14	4.62	20.32
1963	[3]4.40	[3]19.13	(4)	(4)	2.17	9.43	4.39	19.09	8.64	37.57	4.55	19.78
1964	[3]4.46	[3]19.11	(4)	(4)	2.14	9.17	4.45	19.07	8.93	38.26	4.60	19.71
1965	[3]4.45	[3]18.71	(4)	(4)	2.13	8.96	4.44	18.67	8.51	35.79	4.55	19.13
1966	[3]4.56	[3]18.64	(4)	(4)	1.98	8.09	4.54	18.56	8.08	33.03	4.62	18.89
1967	[3]4.64	[3]18.41	(4)	(4)	1.92	7.62	4.62	18.33	8.15	32.33	4.69	18.60
1968	[3]4.70	[3]17.87	(4)	(4)	1.79	6.81	4.67	17.76	8.78	33.38	4.75	18.06
1969	[3]5.02	[3]18.19	(4)	(4)	1.86	6.74	4.99	18.09	9.91	35.92	5.08	18.41
1970	[3]6.30	[3]21.68	(4)	(4)	1.86	6.40	6.26	21.54	11.03	37.96	6.34	21.82
1971	[3]7.13	[3]23.36	(4)	(4)	1.93	6.32	7.07	23.17	12.08	39.58	7.15	23.43
1972	[3]7.78	[3]24.45	(4)	(4)	2.04	6.41	7.66	24.07	12.40	38.97	7.72	24.26
1973	[3]8.71	[3]25.92	(4)	(4)	2.09	6.22	8.53	25.39	13.65	40.63	8.59	25.57
1974	[3]16.01	[3]43.72	(4)	(4)	2.19	5.98	15.75	43.01	22.19	60.60	15.82	43.20
1975	[3]19.79	[3]49.44	(4)	(4)	3.17	7.92	19.23	48.04	32.26	80.59	19.35	48.34
1976	[3]20.11	[3]47.54	(4)	(4)	3.74	8.84	19.43	45.93	33.92	80.19	19.56	46.24
1977	[3]20.59	[3]45.74	(4)	(4)	4.03	8.95	19.82	44.02	34.86	77.43	19.95	44.31
1978	[3]22.64	[3]46.94	(4)	(4)	5.68	11.78	21.76	45.12	35.25	73.09	21.86	45.32
1979	[3]27.31	[3]52.27	9.55	18.28	6.48	12.40	23.66	45.28	41.06	78.58	23.75	45.45
1980	29.17	51.14	11.08	19.42	W	W	24.52	42.99	42.51	74.53	24.65	43.22
1981	31.51	50.52	12.18	19.53	W	W	26.29	42.15	44.28	71.00	26.40	42.33
1982	32.15	48.53	13.37	20.18	W	W	27.14	40.97	49.85	75.25	27.25	41.13
1983	31.11	45.17	13.03	18.92	W	W	25.85	37.53	52.29	75.91	25.98	37.72
1984	30.63	42.88	12.41	17.37	10.45	14.63	25.51	35.71	48.22	67.50	25.61	35.85
1985	30.78	41.77	12.57	17.06	10.68	14.49	25.10	34.06	45.80	62.15	25.20	34.20
1986	28.84	38.30	12.26	16.28	10.64	14.13	23.70	31.47	44.12	58.58	23.79	31.59
1987	28.19	36.34	11.32	14.59	10.85	13.99	23.00	29.65	43.65	56.26	23.07	29.74
1988	27.66	34.48	10.45	13.03	10.06	12.54	22.00	27.43	44.16	55.06	22.07	27.52
1989	27.40	32.91	10.16	12.20	9.91	11.90	21.76	26.13	42.93	51.56	21.82	26.20
1990	27.43	31.71	9.70	11.21	10.13	11.71	21.71	25.10	39.40	45.54	21.76	25.15
1991	27.49	30.66	9.68	10.80	10.89	12.15	21.45	23.92	36.34	40.53	21.49	23.97
1992	26.78	29.16	9.68	10.54	10.81	11.77	20.99	22.85	34.24	37.28	21.03	22.90
1993	26.15	27.80	9.33	9.92	11.11	11.81	19.79	21.04	32.94	35.02	19.85	21.11
1994	25.68	26.75	8.37	8.72	10.77	11.22	19.34	20.14	36.07	37.57	19.41	20.22
1995	25.56	26.06	8.10	8.26	10.83	11.04	18.74	19.10	39.78	40.55	18.83	19.19
1996	25.17	25.17	7.87	7.87	10.92	10.92	18.42	18.42	36.78	36.78	18.50	18.50
1997	24.64	[R]24.17	7.42	7.28	10.91	[R]10.70	18.07	[R]17.72	35.12	[R]34.45	18.14	[R]17.79
1998	24.87	[R]24.09	6.96	[R]6.74	11.08	[R]10.73	17.55	[R]17.00	42.91	[R]41.57	17.67	[R]17.12
1999[P]	23.88	22.79	7.02	6.70	11.04	10.54	16.64	15.88	38.94	37.17	16.76	16.00

[1]Subtotal of bituminous coal, subbituminous coal, and lignite.
[2]In chained (1996) dollars, calculated by using gross domestic product implicit price deflators.
[3]Includes subbituminous coal.
[4]Included in bituminous coal.
R=Revised. P=Preliminary. W=Withheld to avoid disclosure of individual company data.
Note: Prices are free-on-board (f.o.b.) mine prices. FOB is a sales transaction in which the seller makes the product available at a given port and price and the buyer pays for the transportation and insurance.

SOURCE: "Table 7.8 Coal Prices, 1949–1999 (Dollars per Short Ton)," in *Annual Energy Review* 2000, U.S. Department of Energy, Energy Information Administration, Washington, DC, 2001

An Energy Information Administration document, *The U.S. Coal Industry in the 1990s: Low Prices and Record Production* (R. Bonskowski, 1999) notes that the 1990 Clean Air Act reduced coal production from the high-sulfur areas of the Midwest and northern Appalachia during the 1990s. Low-sulfur replacement coal came mostly from central Appalachia and the West. The only major drawback for western coal is the cost of transporta-

FIGURE 4.9

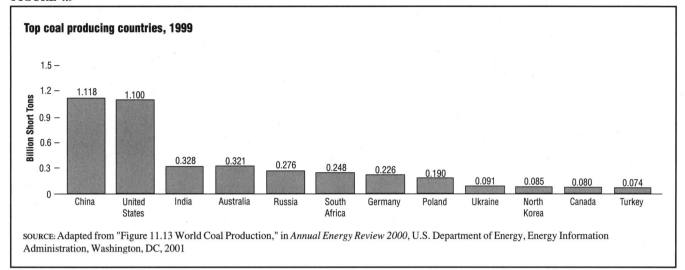

Top coal producing countries, 1999

SOURCE: Adapted from "Figure 11.13 World Coal Production," in *Annual Energy Review 2000*, U.S. Department of Energy, Energy Information Administration, Washington, DC, 2001

tion, but improvements in rail service have reduced this disadvantage.

CLEANER COAL USE. The coal-burning process can be cleaned by physical or chemical methods. Scrubbers, which are a physical method commonly used to reduce sulfur dioxide emissions, filter coal emissions by spraying lime or a calcium compound and water across the emission stream before it leaves the smokestack. The sulfur dioxide bonds to the spray and settles as a mudlike substance that can be pumped out for disposal. Scrubbers, however, are expensive to operate, so particulate collectors are the most common emissions cleaners for coal. While they are cheaper to operate than scrubbers, they are less effective. Cooling towers reduce heat released into the atmosphere and reduce some pollutants. Chemical cleaning, a relatively new technology not yet in widespread use, involves the use of biological or chemical agents to clean emissions.

Under the new environmental regulations of the 1990 Clean Air Act and its amendments, plants with coal-generated boilers must be built to reduce sulfur emissions by 70–90 percent. New, high-sulfur coal electricity plants, designed to meet emission standards, use 30 percent of their construction costs on pollution control equipment and take up to 5 percent of their power output to operate this equipment. Research to lower these costs is important because of the quantity of electricity produced in the United States with coal.

The EIA's *Annual Energy Review 2000* (2001) notes that in 1999 coal-fired electricity plants that had environmental equipment installed had a production capacity of 331 gigawatts (1 gigawatt equals 1,000 megawatts). Of this capacity, 98 percent is generated within plants using particulate collectors, 44 percent in those with cooling towers, and 27 percent in those with scrubbers. The use of scrubbers is projected to increase as new regulations from the 1990 Clean Air Act and its amendments take effect.

COAL EXPORTS

Since 1950 the United States has produced more coal than it has consumed. The excess production has allowed the United States to become a significant exporter of coal to other nations. Exports amounted to 58.5 million short tons in 1999 and in 2000, but those figures are the lowest they have been since 1978. (See Table 4.3.)

In 2000 coal provided 37 percent of all U.S. energy exports. (See Table 4.3.) Europe received 43 percent of U.S. coal exports. The individual countries that bought the most U.S. coal were Canada, Brazil, Japan, Italy, the United Kingdom, and France respectively.

INTERNATIONAL COAL USAGE

The *Annual Energy Review 2000* (EIA, 2001) discloses that world coal production was 4.7 billion short tons in 1999 and accounted for 22 percent of world energy production. China led the world in coal production, mining just over 1.1 billion short tons, followed by the United States at 1.1 billion short tons. (See Figure 4.9.) Other major producers were India, Australia, Russia, South Africa, Germany, and Poland.

World consumption of coal in 1999 totaled 4.7 billion short tons. Besides being the largest producer, China was also the largest consumer of coal in 1999, using nearly 1.1 billion short tons, followed by the United States, which used only slightly less. (See Figure 4.10.) Other major consumers included India, Russia, and Germany.

FUTURE TRENDS IN THE COAL INDUSTRY

The EIA, in its *Annual Energy Outlook 2002* (2001), forecasts that domestic coal production will increase to approximately 1.3 billion short tons by 2015 and increase slightly from there to almost 1.4 billion short tons by 2020. (See Table 4.4.) Domestic consumption is projected

TABLE 4.3

Coal exports by country of destination, 1960–2000

(million short tons)

Year	Canada	Brazil	Belgium and Luxembourg	Denmark	France	Germany[1]	Italy	Netherlands	Spain	United Kingdom	Other	Total (Europe)	Japan	Other	Total
1960	12.8	1.1	1.1	0.1	0.8	4.6	4.9	2.8	0.3	0.0	2.4	17.1	5.6	1.3	38.0
1961	12.1	1.0	1.0	0.1	0.7	4.3	4.8	2.6	0.2	0.0	2.0	15.7	6.6	1.0	36.4
1962	12.3	1.3	1.3	(s)	0.9	5.1	6.0	3.3	0.8	(s)	1.8	19.1	6.5	1.0	40.2
1963	14.6	1.2	2.7	(s)	2.7	5.6	7.9	5.0	1.5	0.0	2.4	27.7	6.1	0.9	50.4
1964	14.8	1.1	2.3	(s)	2.2	5.2	8.1	4.2	1.4	0.0	2.6	26.0	6.5	1.1	49.5
1965	16.3	1.2	2.2	(s)	2.1	4.7	9.0	3.4	1.4	(s)	2.3	25.1	7.5	0.9	51.0
1966	16.5	1.7	1.8	(s)	1.6	4.9	7.8	3.2	1.2	(s)	2.5	23.1	7.8	1.0	50.1
1967	15.8	1.7	1.4	0.0	2.1	4.7	5.9	2.2	1.0	0.0	2.1	19.4	12.2	1.0	50.1
1968	17.1	1.8	1.1	0.0	1.5	3.8	4.3	1.5	1.5	0.0	1.9	15.5	15.8	0.9	51.2
1969	17.3	1.8	0.9	0.0	2.3	3.5	3.7	1.6	1.8	0.0	1.3	15.2	21.4	1.2	56.9
1970	19.1	2.0	1.9	0.0	3.6	5.0	4.3	2.1	3.2	(s)	1.8	21.8	27.6	1.2	71.7
1971	18.0	1.9	0.8	0.0	3.2	2.9	2.7	1.6	2.6	1.7	1.1	16.6	19.7	1.1	57.3
1972	18.7	1.9	1.1	0.0	1.7	2.4	3.7	2.3	2.1	2.4	1.1	16.9	18.0	1.2	56.7
1973	16.7	1.6	1.2	0.0	2.0	1.6	3.3	1.8	2.2	0.9	1.3	14.4	19.2	1.6	53.6
1974	14.2	1.3	1.1	0.0	2.7	1.5	3.9	2.6	2.0	1.4	0.9	16.1	27.3	1.8	60.7
1975	17.3	2.0	0.6	0.0	3.6	2.0	4.5	2.1	2.7	1.9	1.6	19.0	25.4	2.6	66.3
1976	16.9	2.2	2.2	(s)	3.5	1.0	4.2	3.5	2.5	0.8	2.1	19.9	18.8	2.1	60.0
1977	17.7	2.3	1.5	0.1	2.1	0.9	4.1	2.0	1.6	0.6	2.1	15.0	15.9	3.5	54.3
1978	15.7	1.5	1.1	0.0	1.7	0.6	3.2	1.1	0.8	0.4	2.2	11.0	10.1	2.5	40.7
1979	19.5	2.8	3.2	0.2	3.9	2.6	5.0	2.0	1.4	1.4	4.4	23.9	15.7	4.1	66.0
1980	17.5	3.3	4.6	1.7	7.8	2.5	7.1	4.7	3.4	4.1	6.0	41.9	23.1	6.0	91.7
1981	18.2	2.7	4.3	3.9	9.7	4.3	10.5	6.8	6.4	2.3	8.8	57.0	25.9	8.7	112.5
1982	18.6	3.1	4.8	2.8	9.0	2.3	11.3	5.9	5.6	2.0	7.6	51.3	25.8	7.5	106.3
1983	17.2	3.6	2.5	1.7	4.2	1.5	8.1	4.2	3.3	1.2	6.4	33.1	17.9	6.1	77.8
1984	20.4	4.7	3.9	0.6	3.8	0.9	7.6	5.5	2.3	2.9	5.3	32.8	16.3	7.2	81.5
1985	16.4	5.9	4.4	2.2	4.5	1.1	10.3	6.3	3.5	2.7	10.3	45.1	15.4	9.9	92.7
1986	14.5	5.7	4.4	2.1	5.4	0.8	10.4	5.6	2.6	2.9	8.4	42.6	11.4	11.4	85.5
1987	16.2	5.8	4.6	0.9	2.9	0.5	9.5	4.1	2.5	2.6	6.6	34.2	11.1	12.3	79.6
1988	19.2	5.3	6.5	2.8	4.3	0.7	11.1	5.1	2.5	3.7	8.5	45.1	14.1	11.3	95.0
1989	16.8	5.7	7.1	3.2	6.5	0.7	11.2	6.1	3.3	4.5	8.9	51.6	13.8	12.9	100.8
1990	15.5	5.8	8.5	3.2	6.9	1.1	11.9	8.4	3.8	5.2	9.5	58.4	13.3	12.7	105.8
1991	11.2	7.1	7.5	4.7	9.5	1.7	11.3	9.6	4.7	6.2	10.4	65.5	12.3	13.0	109.0
1992	15.1	6.4	7.2	3.8	8.1	1.0	9.3	9.1	4.5	5.6	8.5	57.3	12.3	11.4	102.5
1993	8.9	5.2	5.2	0.3	4.0	0.5	6.9	5.6	4.1	4.1	6.9	37.6	11.9	11.0	74.5
1994	9.2	5.5	4.9	0.5	2.9	0.3	7.5	4.9	4.1	3.4	7.3	35.8	10.2	10.7	71.4
1995	9.4	6.4	4.5	2.1	3.7	2.0	9.1	7.3	4.7	4.7	10.7	48.6	11.8	12.4	88.5
1996	12.0	6.5	4.6	1.3	3.9	1.1	9.2	7.1	4.1	6.2	9.8	47.2	10.5	14.2	90.5
1997	15.0	7.5	4.3	0.4	3.4	0.9	7.0	4.8	4.1	7.2	9.2	41.3	8.0	11.8	83.5
1998	20.7	6.5	3.2	0.3	3.2	1.2	5.3	4.5	3.2	5.9	6.9	33.8	7.7	9.4	78.0
1999	19.8	4.4	2.1	0.0	2.5	0.6	4.0	3.4	2.5	3.2	4.3	22.5	5.0	6.7	58.5
2000	18.8	4.5	2.9	0.1	3.0	1.0	3.7	2.6	2.7	3.3	5.7	25.0	4.4	5.8	58.5

[1]Through 1990, the data for Germany are for the former West Germany only. Beginning with 1991, the data for Germany are for the unified Germany, i.e., the former East Germany and West Germany.
(s)=Less than 0.05 million short tons.
Note: Totals may not equal sum of components due to independent rounding.

SOURCE: "Table 7.4 Coal Exports by Country of Destination, 1960–2000 (Million Short Tons)," in *Annual Energy Review 2000*, U.S. Department of Energy, Energy Information Administration, Washington, DC, 2001.

FIGURE 4.10

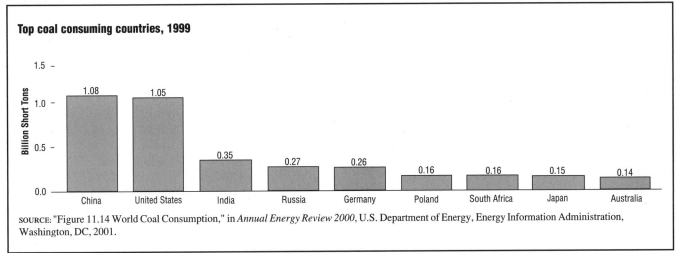

Top coal consuming countries, 1999

SOURCE: "Figure 11.14 World Coal Consumption," in *Annual Energy Review 2000*, U.S. Department of Energy, Energy Information Administration, Washington, DC, 2001.

TABLE 4.4

Comparison of coal forecasts, 2002

(million short tons, except where noted)

Projection	2000	AEO2002			Other forecasts	
		Reference	Low economic growth	High economic growth	DRI-WEFA	Hill & Associates
		2015				
Production	1,084	1,325	1,298	1,366	1,202	1,172
Consumption by sector						
Electricity generation	965	1,183	1,156	1,218	1,048	1,075
Coking plants	29	22	22	22	26	18
Industrial/other	87	89	84	95	85	61
Total	1,081	1,294	1,262	1,335	1,159	1,154
Net coal exports	46	34	39	34	44	18
Minemouth price						
(2000 dollars per short ton)	16.45	13.44	13.17	13.51	NA	9.77[a]
(2000 dollars per million Btu)	0.79	0.66	0.65	0.67	NA	0.46[a]
Average delivered price to electricity generators						
(2000 dollars per short ton)	24.36	20.15	19.91	20.55	26.03	18.23
(2000 dollars per million Btu)	1.20	1.01	1.00	1.03	1.26	0.87
		2020				
Production	1,084	1,397	1,335	1,493	1,208	1,160
Consumption by sector						
Electricity generation	965	1,254	1,198	1,341	1,054	1,071
Coking plants	29	20	20	20	24	15
Industrial/other	87	92	85	101	88	58
Total	1,081	1,365	1,303	1,462	1,166	1,144
Net coal exports	46	35	35	35	42	16
Minemouth price						
(2000 dollars per short ton)	16.45	12.79	12.56	13.23	NA	8.98[a]
(2000 dollars per million Btu)	0.79	0.64	0.62	0.66	NA	0.42[a]
Average delivered price to electricity generators						
(2000 dollars per short ton)	24.36	19.00	18.72	19.75	24.70	17.26
(2000 dollars per million Btu)	1.20	0.97	0.95	1.00	1.20	0.82

[a]In the Hill & Associates forecast, minemouth prices represent an average for domestic steam coal only. Exports and coking coal are not included in the average.
NA = Not available.
Btu = British thermal unit.

SOURCE: "Table 25. Comparison of coal forecasts (million short tons, except where noted)," in *Annual Energy Outlook 2002*, U.S. Department of Energy, Energy Information Administration, Washington, DC, 2001. EIA data sources include *AEO2002, The Outlook for U.S. Steam Coal: Long-Term Forecast to 2020*, from Hill & Associates, and *U.S. Energy Outlook* from DRI-WEFA.

to match production and stabilize as well, reaching nearly 1.3 billion short tons in 2015 and staying at that level through 2020. Electricity generation will still use the majority of coal in 2020 (nearly 1.2 billion short tons), but coal's share in total electricity generation will decrease slightly because of the increased use of cleaner and more efficient natural gas.

As shown in Table 4.4, coal prices are projected to decline from $16.45 per short ton in 2000 to $13.44 per short ton in 2015 and $12.79 per short ton in 2020. These price decreases will likely occur because of improvements in mine productivity and the lower costs of mining in the West. Eastern mines will probably lose market share to less expensive low-sulfur coal from western mines. Also a factor are declining coal transportation and labor costs in the West.

Environmental concerns about acid rain and global warming may continue to grow. The outlook for the U.S. coal industry could be affected by acid rain legislation, the development of clean coal technologies, and, over the longer term, the problem of global warming. Environmental issues will increasingly become worldwide issues. China, with nearly five times the population of the United States and a growing economy, may surpass the United States in carbon emissions by 2020.

The EIA predicts that U.S. coal exports will decline from 10 percent in 2000 to 8 percent by 2020. This decline will likely be the result of a decline in the demand for coal in Europe and the Americas. Also, other countries are expected to reduce costs and gain currency exchange advantages against the U.S. dollar.

CHAPTER 5

NUCLEAR ENERGY

CONQUERING OIL DEPENDENCY?

A good deal of environmental contamination is directly related to the use of fossil fuels. To prevent further damage, many energy experts have turned their attention to other means of energy production. Nuclear energy is currently used for steam generation at electric utilities, for ship propulsion, and for nuclear weapons. Some observers consider nuclear energy an attractive alternative because it does not pollute the atmosphere as do fossil fuels. However, producing nuclear energy depends on uranium, which is a nonrenewable resource. In addition, safe disposal of radioactive waste, which is a by-product of nuclear energy, has proved difficult.

Many citizens do not want nuclear power plants in their neighborhoods, while others oppose nuclear power for broader environmental reasons. In the decades since the first commercial nuclear reactor went into operation in 1956, the nuclear power industry has had a difficult time persuading many Americans of the safety of its enterprise. In the early 1970s most Americans favored the use of nuclear power; in the early 2000s many Americans—and, indeed, many people around the world—oppose building additional nuclear power plants. The 1979 near disaster at Three Mile Island in Pennsylvania and the 1986 catastrophe at Chernobyl in the former Soviet Union greatly increased public concerns about the safety of nuclear power, as have reports of design flaws, cracks, and leaks in other reactors.

Supporters of nuclear power believe that it is as safe as any other form of energy production if monitored correctly. They point to growing concerns about fossil fuel use, including global warming and acid rain, as well as damage caused by mining and transporting fossil fuels. In fact, this growing concern over fossil fuels has led a small number of environmentalists who had previously opposed nuclear power to reconsider their position. Nonetheless, environmental, safety, and economic concerns have restrained growth in the nuclear industry since the mid-

FIGURE 5.1

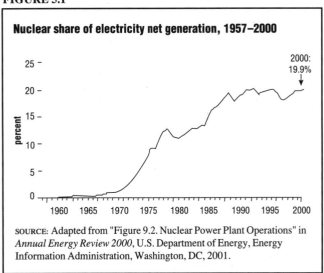

Nuclear share of electricity net generation, 1957–2000

2000: 19.9%

SOURCE: Adapted from "Figure 9.2. Nuclear Power Plant Operations" in *Annual Energy Review 2000*, U.S. Department of Energy, Energy Information Administration, Washington, DC, 2001.

1970s. Unwillingness to commission new nuclear plants is evident in Figure 5.1, which shows a general leveling off of nuclear energy's share of electricity production from about 1988 through 2000.

When it was introduced in the 1950s, nuclear power was presented as the energy of the future, a source so cheap it would eliminate the need for electric meters. Those dreams never came true. Construction and operating costs skyrocketed and far exceeded early estimates. The technology was considerably more complicated than originally thought, and it turned out to be increasingly expensive to build and run reactors that would meet Nuclear Regulatory Commission (NRC) standards.

Nevertheless, in January 2000 the United States entered into a collaboration with international partners on nuclear energy. Formally chartered in 2001, the Generation IV International Forum (GIF) is a group of leading nuclear nations that agree that nuclear energy is important to future world energy security and economic prosperity.

FIGURE 5.2

The nuclear fuel cycle

The commercial nuclear fuel cycle includes activities for preparing and using reactor fuel and for managing spent fuel and other radioactive wastes produced in the process. It was originally intended that spent fuel be stored for 6 months in water-filled basins at reactors sites to dissipate thermal heat and allow decay of short-lived fission products. The spent fuel would then be reprocessed and the resultant liquid high-level waste solidified and disposed of in a federal repository.

SOURCE: Council on Environmental Quality

FIGURE 5.3

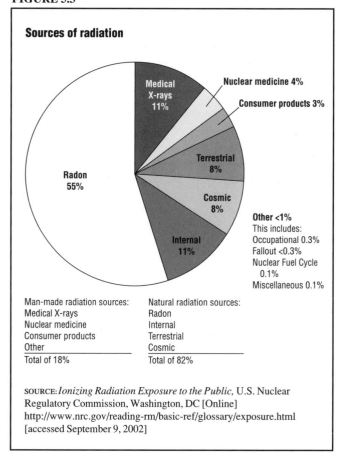

Sources of radiation

Man-made radiation sources:
Medical X-rays
Nuclear medicine
Consumer products
Other
Total of 18%

Natural radiation sources:
Radon
Internal
Terrestrial
Cosmic
Total of 82%

SOURCE: *Ionizing Radiation Exposure to the Public,* U.S. Nuclear Regulatory Commission, Washington, DC [Online] http://www.nrc.gov/reading-rm/basic-ref/glossary/exposure.html [accessed September 9, 2002]

1970s and 1980s; and (c) Generation III advanced light-water reactors built in the 1990s, primarily in East Asia, to meet that region's expanding electricity needs.

HOW NUCLEAR ENERGY WORKS

Figure 5.2 shows the nuclear fuel cycle. In a nuclear power plant, fuel (uranium in the United States) in the reactor generates a nuclear reaction (fission) that produces heat. In a pressurized water reactor, the heat from the reaction is carried away by water under high pressure, which heats a second water stream, producing steam. The steam runs through a turbine (similar to a jet engine), making it and the attached electrical generator spin, which produces electricity. The steam is then cooled and recirculated. The large cooling towers associated with nuclear plants are used to cool the steam after it has run through the turbines. A boiling water reactor works much the same way, except that the water surrounding the core boils and directly produces the steam, which is then piped to the turbine generator.

The key problems in operating a nuclear power reactor include finding material (uranium 235) that will sustain a chain reaction, maintaining the reaction at a level that yields heat but does not escalate out of control and explode, and coping with the radiation produced by the chain reaction.

These countries are dedicated to joint development of the "next generation" of nuclear energy systems. Members of GIF are the United States, Argentina, Brazil, Canada, France, Japan, the Republic of Korea, South Africa, Switzerland, and the United Kingdom.

The GIF countries have selected six nuclear power technologies to develop for the future. These "Generation IV" nuclear energy systems would follow the three other periods of nuclear reactor development: (a) Generation I experimental reactors developed in the 1950s and 1960s; (b) Generation II large, central-station nuclear power reactors, such as the 104 plants still operating in the United States, built in the

FIGURE 5.4

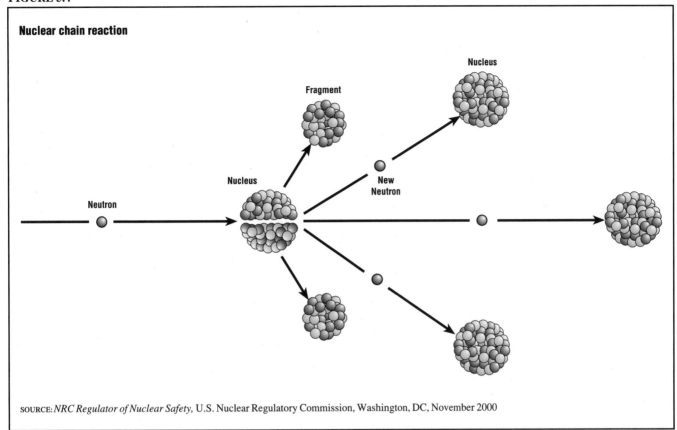

Nuclear chain reaction

Nucleus

Fragment

Nucleus

New
Neutron

Nucleus

Neutron

SOURCE: *NRC Regulator of Nuclear Safety,* U.S. Nuclear Regulatory Commission, Washington, DC, November 2000

Radioactivity

Radioactivity is the spontaneous emission of energy and/or high-energy particles from the nucleus of an atom. One type of radioactivity is produced naturally and is emitted by radioactive isotopes (or radioisotopes), such as radioactive carbon (carbon 14) and radioactive hydrogen (H-3, or tritium). The energy and high-energy particles radioactive isotopes emit include alpha rays, beta rays, and gamma rays.

Isotopes are atoms of an element that have the usual number of protons but different numbers of neutrons in their nuclei. For example, the nucleus of the element carbon comprises 12 protons and 12 neutrons. One isotope of carbon, C-14, has 12 protons and 14 neutrons in its nucleus.

Radioisotopes (such as C-14) are unstable isotopes and their nuclei decay, or break apart, at a steady rate. Decaying radioisotopes produce other isotopes as they emit energy and/or high-energy particles. If the newly formed nuclei are radioactive too, they emit radiation and change into other nuclei. The final products in this chain are stable, nonradioactive nuclei.

Radioisotopes reach our bodies daily, emitted from sources in outer space, and from rocks and soil on earth. Radioisotopes are also used in medicine and provide useful diagnostic tools. Figure 5.3 shows the sources of radiation.

Decades after the discovery of radiation at the turn of the century by Antoine Henri Becquerel, Marie Curie, and Pierre Curie, other scientists determined that they could unleash energy by "artificially" breaking apart atomic nuclei. Such a process is called nuclear fission. Scientists learned that they could produce the most energy by bombarding the nuclei of an isotope of uranium called uranium 235 (U-235). The fission of U-235 releases several neutrons, which can penetrate other U-235 nuclei. In this way, the fission of a single U-235 atom could begin a cascading chain of nuclear reactions, as shown in Figure 5.4. If this series of reactions is regulated to occur slowly, as it is in nuclear power plants, the energy emitted can be captured for a variety of uses, such as generating electricity. If this series of reactions is allowed to occur all at once, as in a nuclear (atomic) bomb, the energy emitted is explosive. (Plutonium 239 can also be used to generate a chain reaction similar to that of U-235.)

Mining Nuclear Fuel

In the United States, U-235, found in the form of ore, is used as nuclear fuel. The majority of uranium in the United States is found in two states: Wyoming and New Mexico. Ore that contains uranium is first located by geological methods such as drilling. Uranium-bearing ores are mined by methods similar to those used for other metal ores. However, uranium mining has the unique

TABLE 5.1

Nuclear power plant operations, 1957–2000

Year	Nuclear electricity net generation Billion kilowatthours	Nuclear share of electricity net generation Percent	Net summer capability of operable units[1] Million kilowatts	Capacity factor Percent
1957	(s)	(s)	0.1	NA
1958	0.2	(s)	0.1	NA
1959	0.2	(s)	0.1	NA
1960	0.5	0.1	0.4	NA
1961	1.7	0.2	0.4	NA
1962	2.3	0.3	0.7	NA
1963	3.2	0.4	0.8	NA
1964	3.3	0.3	0.8	NA
1965	3.7	0.3	0.8	NA
1966	5.5	0.5	1.7	NA
1967	7.7	0.6	2.7	NA
1968	12.5	0.9	2.7	NA
1969	13.9	1.0	4.4	NA
1970	21.8	1.4	7.0	NA
1971	38.1	2.4	9.0	NA
1972	54.1	3.1	14.5	NA
1973	83.5	4.5	22.7	53.5
1974	114.0	6.1	31.9	47.8
1975	172.5	9.0	37.3	55.9
1976	191.1	9.4	43.8	54.7
1977	250.9	11.8	46.3	63.3
1978	276.4	12.5	50.8	64.5
1979	255.2	11.4	49.7	58.4
1980	251.1	11.0	51.8	56.3
1981	272.7	11.9	56.0	58.2
1982	282.8	12.6	60.0	56.6
1983	293.7	12.7	63.0	54.4
1984	327.6	13.6	69.7	56.3
1985	383.7	15.5	79.4	58.0
1986	414.0	16.6	85.2	56.9
1987	455.3	17.7	93.6	57.4
1988	527.0	19.5	94.7	63.5
1989	[2]529.4	[2]17.8	[2]98.2	[2]62.2
1990	577.0	19.1	99.6	66.0
1991	612.6	19.9	99.6	70.2
1992	618.8	20.1	99.0	70.9
1993	610.4	19.1	99.1	70.5
1994	640.5	19.7	99.1	73.8
1995	673.4	20.1	99.5	77.4
1996	674.7	19.6	100.8	76.2
1997	628.6	18.0	99.7	71.1
1998	673.7	18.6	97.1	78.2
1999	R728.3	R19.7	97.5	R85.3
2000P	753.9	19.9	97.4	88.1

[1]At end of year.
[2]Beginning in 1989, includes nonutility facilities.
R=Revised. P=Preliminary. NA=Not available. (s)=Less than 0.05 billion kilowatthours or less than 0.05 percent.
Note: The performance data shown in this table are based on a universe of reactor units that differs in some respects from the reactor universe used to profile the nuclear power industry, especially in the years prior to 1973.

SOURCE: "Table 9.2. Nuclear Power Plant Operations, 1957–2000," in *Annual Energy Review 2000*, U.S. Department of Energy, Energy Information Administration, Washington, DC, 2001.

danger of exposure to radioactivity. Uranium atoms split by themselves at a slow rate, causing radioactive substances such as radon to accumulate slowly in the deposits.

After it is mined, uranium must be concentrated, because uranium ore generally contains only 0.1 percent uranium metal by weight. To concentrate the uranium, it goes through a process called milling. (See Figure 5.2.) In milling, the uranium is first crushed, and then various chemicals are poured slowly through the crushed ore to dissolve out the uranium. The uranium is then precipitated from this chemical solution. The resulting material, called yellowcake because of its color, is 85 percent pure uranium by weight. However, this uranium is 99.3 percent nonfissionable U-238 and only 0.7 percent fissionable U-235. Other processes result in enriched uranium, which is uranium that has a higher percentage of fissionable uranium than does yellowcake.

Nuclear fuel processing plants shape enriched uranium into very small cylinders, called pellets. The small pellets are less than one-half inch in diameter, but each one can produce as much energy as 120 gallons of oil. The pellets are stacked in tubes about 12 feet long, called rods. Many rods are bundled together in assemblies, and hundreds of these assemblies make up the core of a nuclear reactor.

DOMESTIC NUCLEAR ENERGY PRODUCTION

The percentage of U.S. electricity supplied by nuclear power grew considerably during the 1970s and 1980s before leveling off in the 1990s. (See Table 5.1.) In 1973 nuclear power supplied only 4.5 percent of the total electricity generated in the United States. By 2000 nuclear power's 19.9 percent share of electricity was down only slightly from the 1992 and 1995 highs of 20.1 percent and provided an all-time high of 753.9 billion kilowatt-hours of electricity.

This increase in production of electricity was achieved largely through an increase in average capacity factor at American nuclear power plants. The capacity factor is the proportion of electricity produced to what could have been produced at full-power operation. Better training for operators, longer operating cycles between refueling, and control-system improvements contributed to this improved plant performance. According to the Energy Information Administration (EIA) of the U.S. Department of Energy (DOE), in 2000 U.S. nuclear power plants achieved a record overall average capacity factor of 88.1 percent, up from 58.4 percent in 1979, the year of the Three Mile Island accident. (See Table 5.1.)

In 2000, 104 nuclear reactors (see Figure 5.5) were operating in 32 states (see Figure 5.6), with a total net generation of 754 billion kilowatt-hours of energy (see Table 5.1). Most of these reactors are located east of the Mississippi River, where the demand for electricity is high. The number of plants operating is lower than the 135 reactors that were either planned, in construction, or in operation in 1974.

Since 1978 no new nuclear power plants have been ordered, and many have closed. (See Table 5.2.) Several

FIGURE 5.5

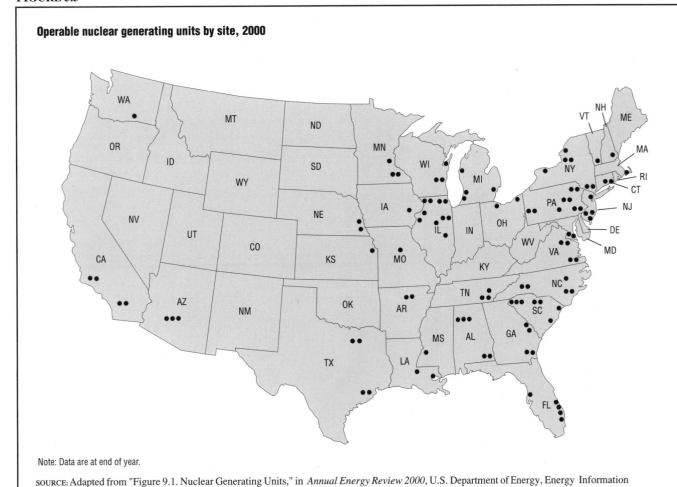

Operable nuclear generating units by site, 2000

Note: Data are at end of year.

SOURCE: Adapted from "Figure 9.1. Nuclear Generating Units," in *Annual Energy Review 2000*, U.S. Department of Energy, Energy Information Administration, Washington, DC, 2001. EIA data sources include various U.S. government reports as well as information from *Historical Profile of U.S. Nuclear Power Development,* 1988, from the Nuclear Energy Institute.

factors have contributed to the slowdown in U.S. nuclear reactor construction. The demand for electricity has grown more slowly than expected. Overall costs have increased as a result of more expensive financing, partly influenced by longer delays for licensing. Expenses have also increased because of regulations that were instituted as a result of the Three Mile Island incident. Operating costs have been higher than expected, as well. Originally, it was projected that plants could run almost 90 percent of the time, with brief pauses for refueling, but for many years this was not the case. For example, Indian Point 3, a nuclear power plant in New York, has been in operation only about 50 percent of the time.

OUTLOOK FOR DOMESTIC NUCLEAR ENERGY

In its *Annual Energy Outlook 2002* (2001), the EIA predicts that nuclear energy's contribution to U.S. electricity production will decline 10 percent over the next 20 years, as aging nuclear plants are shut down. Electricity needs are expected to be met by coal and natural gas sources, which are more economical.

FIGURE 5.6

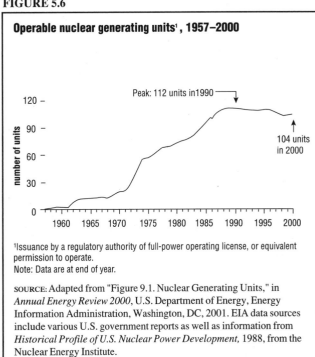

Operable nuclear generating units[1] , 1957–2000

[1]Issuance by a regulatory authority of full-power operating license, or equivalent permission to operate.
Note: Data are at end of year.

SOURCE: Adapted from "Figure 9.1. Nuclear Generating Units," in *Annual Energy Review 2000*, U.S. Department of Energy, Energy Information Administration, Washington, DC, 2001. EIA data sources include various U.S. government reports as well as information from *Historical Profile of U.S. Nuclear Power Development,* 1988, from the Nuclear Energy Institute.

TABLE 5.2

Nuclear generating units, 1953–2000

Year	Orders[1]	Construction permits[2]	Low-power operating licenses[3]	New operable units[4]	Shutdowns[5]	Total operable units[6]	Cancellations[7]	Cumulative cancellations
1953	1	0	0	0	0	0	0	0
1954	0	0	0	0	0	0	0	0
1955	3	1	0	0	0	0	0	0
1956	1	3	0	0	0	0	0	0
1957	2	1	1	1	0	1	0	0
1958	4	0	0	0	0	1	0	0
1959	4	3	1	1	0	2	0	0
1960	1	7	1	1	0	3	0	0
1961	0	0	0	0	0	3	0	0
1962	2	1	7	6	0	9	0	0
1963	4	1	3	2	0	11	0	0
1964	0	3	2	3	1	13	0	0
1965	7	1	0	0	0	13	0	0
1966	20	5	1	2	1	14	0	0
1967	29	14	3	3	2	15	0	0
1968	16	23	0	0	2	13	0	0
1969	9	7	4	4	0	17	0	0
1970	14	10	4	3	0	20	0	0
1971	21	4	5	2	0	22	0	0
1972	38	8	6	6	1	27	7	7
1973	42	14	12	15	0	42	0	7
1974	28	23	14	15	2	55	9	16
1975	4	9	3	2	0	57	13	29
1976	3	9	7	7	1	63	1	30
1977	4	15	4	4	0	67	10	40
1978	2	13	3	4	1	70	13	53
1979	0	2	0	0	1	69	6	59
1980	0	0	5	2	0	71	15	74
1981	0	0	3	4	0	75	9	83
1982	0	0	6	4	1	78	18	101
1983	0	0	3	3	0	81	6	107
1984	0	0	7	6	0	87	6	113
1985	0	0	7	9	0	96	2	115
1986	0	0	7	5	0	101	2	117
1987	0	0	6	8	2	107	0	117
1988	0	0	1	2	0	109	3	120
1989	0	0	3	4	2	111	0	120
1990	0	0	1	2	1	112	1	121
1991	0	0	0	0	1	111	0	121
1992	0	0	0	0	2	109	0	121
1993	0	0	1	1	0	110	0	121
1994	0	0	0	0	1	109	1	122
1995	0	0	1	0	0	109	2	124
1996	0	0	0	1	1	109	0	124
1997	0	0	0	0	2	107	0	124
1998	0	0	0	0	3	104	0	124
1999	0	0	0	0	0	104	0	124
2000	0	0	0	0	0	104	0	124

[1]Placement of an order by a utility or government agency for a nuclear steam supply system.
[2]Issuance by regulatory authority of a permit, or equivalent permission, to begin construction. Numbers reflect permits issued in a given year, not extant permits.
[3]Issuance by regulatory authority of license, or equivalent permission, to conduct testing but not to operate at full power.
[4]Issuance by regulatory authority of full-power operating license, or equivalent permission. Units generally did not begin immediate operation.
[5]Ceased operation permanently.
[6]Total of units holding full-power licenses, or equivalent permission to operate, at the end of the year.
[7]Cancellation by utilities of ordered units. Does not include three units (Bellefonte 1 and 2 and Watts Bar 2) where construction has been stopped indefinitely.
Note: Data are at end of year.

SOURCE: "Table 9.1. Nuclear Generating Units, 1953–2000," in *Annual Energy Review 2000*, U.S. Department of Energy, Energy Information Administration, Washington, DC, 2001. EIA data sources include various U.S. government reports as well as information from *Historical Profile of U.S. Nuclear Power Development,* 1988, from the Nuclear Energy Institute.

Nuclear energy's short-term future may vary depending on several factors. Nuclear plants could be granted extensions for their operating licenses if aging problems are not severe and if safety improvements are not too expensive. Nuclear energy capacity could decline less than 10 percent if performance improvements are gained, if the nuclear waste-disposal problem is alleviated, or if stricter emissions regulations for fossil fuels are enacted. However, if nuclear energy operating costs are higher than expected because of severe degradation of the plants or from waste disposal problems, then nuclear power generation will likely decline more than 10 percent.

INTERNATIONAL PRODUCTION

In 1999 nuclear power provided about 17 percent of the world's electricity and 7 percent of total energy. Although the United States is the largest producer of nuclear power, it trails other Western countries in the proportion of national electrical production generated by nuclear power. That is primarily because oil, natural gas, and coal have been more accessible in the United States than in other countries.

In 1999 the United States led the world in nuclear power generation with 728.3 billion kilowatt-hours, followed by France's 375.1 billion kilowatt-hours and Japan's 308.7 billion kilowatt-hours. (See Table 5.3.) These three countries together generated 59 percent of the world's nuclear electric power. France had by far the highest proportion (approximately 75 percent) of its electrical power produced by nuclear energy, followed by Belgium (58 percent), Sweden (45 percent), and Switzerland (35 percent). Japan produces 30 percent of its electricity by nuclear power generation, and the United States only 20 percent.

The EIA reported that as of January 2002, there were 438 nuclear power plants in operation worldwide and 30 were under construction, including 8 in China, 4 in Korea, 4 in the Ukraine, and 3 in Japan. Other countries building nuclear power plants in 2002 were India, Iran, Russia, Slovakia, Argentina, the Czech Republic, and Romania. In western Europe, Finland decided in 2002 to build a nuclear power plant to add to the four facilities it currently has operating. Two new plants began operation in France in 2002.

Worldwide, the EIA projects that nuclear generating capacity will increase slightly overall, from 350 gigawatts in 2000 to 359 gigawatts in 2020. Totals for individual countries will increase for some and decrease for others. Countries whose capacities are projected to increase significantly from 2000 to 2020 are:

- Canada, from 9,998 net megawatts (electric) to 13,596 net megawatts

- China, from 2,167 to 18,652 net megawatts

- India, from 2,301 to 7,571 net megawatts

- Japan, from 43,691 to 56,637 net megawatts

- South Korea, from 12,990 to 22,125 net megawatts

- Iran, from 0 to 2,146 net megawatts

Countries whose capacities are projected to decrease significantly from 2000 to 2020 are:

- The United States, from 97,478 net megawatts to 71,581 net megawatts

- Germany, from 21,122 to 13,134 net megawatts

- The United Kingdom, from 12,498 to 5,333 net megawatts

- Russia, from 19,843 to 13,097 net megawatts

TABLE 5.3

World net nuclear electric power generation, 1980, 1998, and 1999
(billion kilowatthours)

Region and country	Nuclear electric power		
	1980	1998	1999P
North America	**287.0**	**R750.2**	**807.6**
Canada	35.9	R67.7	69.8
Mexico	0.0	8.8	9.5
United States	251.1	673.7	728.3
Other	0.0	0.0	0.0
Central and South America	**2.2**	**10.3**	**10.5**
Argentina	2.2	7.1	6.7
Brazil	0.0	3.1	3.8
Paraguay	0.0	0.0	0.0
Venezuela	0.0	0.0	0.0
Other	0.0	0.0	0.0
Western Europe	**219.2**	**R841.0**	**850.2**
Belgium	11.9	43.9	46.6
Finland	6.6	20.8	21.8
France	63.4	R368.6	375.1
Germany	55.6	R153.6	161.0
Italy	2.1	0.0	0.0
Netherlands	3.9	3.6	3.6
Norway	0.0	0.0	0.0
Spain	5.2	56.0	55.9
Sweden	25.3	R69.9	66.6
Switzerland	12.9	24.5	23.7
Turkey	0.0	0.0	0.0
United Kingdom	32.3	R95.1	91.5
Other	R0.0	R 5.0	4.5
Eastern Europe and Former U.S.S.R.	**83.2**	**239.3**	**245.7**
Czech Republic	—	12.5	12.7
Kazakhstan	—	0.1	0.1
Poland	0.0	0.0	0.0
Romania	0.0	4.9	4.8
Russia	—	98.3	110.9
Ukraine	—	70.6	67.4
Other	83.2	52.8	49.8
Middle East	**0.0**	**0.0**	**0.0**
Iran	0.0	0.0	0.0
Saudi Arabia	0.0	0.0	0.0
Other	0.0	0.0	0.0
Africa	**0.0**	**13.6**	**12.8**
Egypt	0.0	0.0	0.0
South Africa	0.0	13.6	12.8
Other	0.0	0.0	0.0
Far East and Oceania	**92.7**	**R460.8**	**469.1**
Australia	0.0	0.0	0.0
China	0.0	13.5	14.1
India	3.0	10.6	11.5
Indonesia	0.0	0.0	0.0
Japan	78.6	R315.7	308.7
South Korea	3.3	85.2	97.9
Taiwan	7.8	R35.4	36.9
Thailand	0.0	0.0	0.0
Other	0.0	0.4	0.1
World	**684.4**	**R2,315.3**	**2,396.0**

R=Revised. P=Preliminary. — = Not applicable.
Notes: Data include both electric utility and nonutility sources. Totals may not equal sum of components due to independent rounding.

SOURCE: Adapted from "Table 11.15. World Net Generation of Electricity by Type, 1980, 1998, and 1999," in *Annual Energy Review 2000*, U.S. Department of Energy, Energy Information Administration, Washington, DC, 2001

Some countries are planning to cut capacity radically or phase out nuclear power altogether. Sweden has legislated a nuclear phaseout by 2010 (although it is not clear that this is an obtainable goal, since nearly half of Sweden's electricity comes from nuclear power). In 2001 the EIA projected that Sweden's nuclear generating capacity would shift from 9,432 net megawatts in 2000 to 6,077 in 2020. The Netherlands and Germany have strong conservation policies and well-organized antinuclear movements. The Netherlands is projected to phase out nuclear power altogether by 2020, although its capacity in 2000 was only 449 net megawatts. Germany's nuclear generating capacity is projected to drop by more than one-third from 2000 to 2020. While total worldwide nuclear generating capacity is expected to increase from 2000 to 2020, its share of the world electricity supply is expected to drop from 16 percent in 1999 to 12 percent by 2020, mainly because few new plants will be built and some old plants will be retired.

International Agreement on Safety

In September 1994, 40 nations, including the United States, signed the International Convention on Nuclear Safety, an agreement that requires them to shut down nuclear power plants if necessary safety measures cannot be guaranteed. The agreement applies to land-based civil nuclear power plants and seeks to avert accidents like the 1986 explosion at Chernobyl in the former Soviet Union, the world's worst nuclear disaster. Ukraine, which inherited the Chernobyl plant after the collapse of the Soviet Union, signed the agreement. Signers must submit reports on atomic installations and, if necessary, make improvements to upgrade safety at the sites. Neighboring countries may call for an urgent study if they are concerned about a reactor's safety and the potential fallout that could affect their own population or crops.

POWER PLANT AGING

Like other types of power plants, nuclear power plants have a lifespan after which they must be retired because the cost of running them becomes too high. The current generation of nuclear power plants was designed to last about 30 years. Some plants showed serious levels of deterioration after as few as 15 years. However, some have lived well beyond 30-year expectations. Whenever the life of a power plant—whether coal, natural gas, or nuclear—ends, it must be demolished and cleaned up so that the site can be used again.

Decommissioning Nuclear Power Plants

Eventually, the more than 400 nuclear plants now operating worldwide will need to be retired. There are three methods of retiring, or decommissioning, a reactor: safe enclosure (mothballing), entombment, and immediate dismantling.

Safe enclosure involves removing the fuel from the plant, monitoring any radioactive contamination (which is usually very low or nonexistent), and guarding the structure to prevent anyone from entering until eventual dismantling and decontamination activities occur. Entombment, which was used at Chernobyl, involves permanently encasing the structure in a long-lived material such as concrete. This procedure allows the radioactive material to remain safely on-site but at one location. Immediate dismantling involves decontaminating and tearing down the facility within a few months or years. This process is initially more expensive than the other options but removes the long-term costs of monitoring both the structure and the radiation levels. It also frees the site for other uses, including even the construction of another nuclear power plant. Nuclear power officials are also seeking alternative uses for nuclear shells, including the conversion of old nuclear plants to gas-fired plants.

Paying for the closing, decontaminating, or dismantling of nuclear plants has become an issue of intense public debate. The early closing of several plants owing to safety concerns and poor economic performance has raised the question of how the costs of decommissioning can be covered. Industry reports contend that the cost of decommissioning retired plants and handling radioactive wastes will continue to escalate, causing serious financial problems for electric utilities. Most electric utilities have concluded that nuclear power is no longer competitive with other power sources. Not only coal plants but also new technologies, such as wind turbines and geothermal energy, are less expensive than nuclear energy plants when all factors are considered.

NUCLEAR SAFETY PROBLEMS

Problems have plagued the nuclear industry almost from the beginning. Some plant-site selections have been considered questionable, especially those near earthquake fault lines. Internal quality control has often been lax. The NRC has come under fire for failing to follow up adequately on the observations of so-called "whistle-blowers," or those who report problems in plant construction or operation. Several major accidents have occurred worldwide since the late 1970s.

In the United States

THREE MILE ISLAND. On March 28, 1979, the Three Mile Island nuclear facility near Harrisburg, Pennsylvania, was the site of the worst nuclear accident in American history. Information released several years after the accident revealed that the plant came much closer to meltdown than either the NRC or the industry had previously indicated. Temperatures inside the reactor were first said to have been 3,500 degrees Fahrenheit but are now known to have reached at least 4,800 degrees. The temperature needed to melt uranium dioxide fuel is 5,080 degrees Fahrenheit. When meltdown occurs, an uncontrolled explosion may result, unlike the controlled nuclear reaction of normal operation.

The emergency core cooling system at Three Mile Island was designed to dump water on the hot core and spray water into the reactor building to stop the production of steam. But during the accident, the valves leading to the emergency water pumps closed. Another valve was stuck in the open position, drawing water away from the core, which then became partially uncovered and began to melt. The emergency core cooling system then began drawing water out of the basement supply and reusing it, contaminating the reactor pump, although limiting the radiation contamination to the interior of the building.

Although safety systems at the Three Mile Island facility performed well and the radiation leak was relatively small, the Three Mile incident was sensationalized in the media. Antinuclear proponents urged residents to evacuate, adding to stress levels and panic. In the end, this nuclear accident led to no deaths or injuries to plant workers or members of the nearby community. On average, area residents were exposed to less radiation than that of a chest X ray. Nevertheless, the Three Mile Island incident raised concerns about nuclear safety, which resulted in safety enhancements in the nuclear power industry and at the NRC. Antinuclear sentiment was fueled as well, heightening Americans' criticisms of nuclear power as an energy source.

The nuclear reactor in Unit 2 at Three Mile Island has been in monitored storage since it underwent cleanup. Operation of the reactor in Unit 1 resumed in 1985.

PEACH BOTTOM. In 1987 the NRC shut down the Peach Bottom nuclear plant in Delta, Pennsylvania, because control-room operators were found sleeping on duty. This was the first plant shut down solely because of operator violations and misconduct. Many industry observers note that attempts to improve nuclear plant safety focus mainly on technical design improvements and do not recognize "people problems," which are just as serious as mechanical problems. A nuclear analyst and spokesperson for Public Citizen, Ralph Nader's consumer advocacy group, observed that while the focus of the NRC is primarily technological, human error has been responsible for almost all nuclear accidents. Even supporters of nuclear power recognize that many of the jobs in question can be tedious, although they take a great deal of training to master.

In the Former Soviet Union

CHERNOBYL. On April 26, 1986, the most serious nuclear accident ever occurred at Chernobyl, a four-reactor nuclear plant complex located in the former Soviet Union (now Ukraine) near Kiev. At least 31 people died, and hundreds were injured in the explosion. About 500 people were hospitalized, and medical experts estimate that 6,000 to 24,000 cancer-related deaths have occurred over the intervening years as a result of the released radiation.

The cleanup was one of the biggest projects of its kind ever undertaken. Helicopters dropped 5,000 tons of limestone, sand, clay, lead, and boron on the smoldering reactor to stop the radiation leakage and reduce the heat. Workers built a giant steel and cement sarcophagus (stone coffin) to enshrine the remains of the reactor and contain the radioactive waste. Approximately 135,000 people were evacuated from a 300-square-mile area around the power station. Topsoil had to be removed in a 19-mile area and buried as nuclear waste. Buildings were washed down and the newly contaminated water and soil were carted away and buried. Agricultural products from areas nearby were declared unmarketable throughout Europe. Many of the 600,000 people involved in the immediate cleanup have suffered long-term effects of radiation exposure.

Some people at Chernobyl received 400 rems of radiation immediately following the explosion. (A rem is a standard measure of the whole-body dose of radiation. Under normal conditions, a person receives about one-tenth of a rem annually.) With a dose of 25 rems, a person's blood begins to change. The DNA is damaged, preventing the bone marrow from producing red and white blood cells. Sickness starts at 100 rems and severe sickness at 200 rems. Death of half the population occurs at 400 rems, and death within a week can be expected for anyone exposed to 600 rems. One estimate is that 17.5 million people, including 2.5 million children under seven years old, have had some significant exposure to radiation from Chernobyl.

In 1995 the United Nations reported that the prevalence of illnesses of all kinds was 30 percent above normal in the Ukraine; incidences of depression, alcoholism, and divorce were on the rise. Although the Chernobyl reactor is in the Ukraine, the neighboring nation of Belarus (with a population of 10 million people) suffered more human and ecological damage from fallout because of the prevailing winds. Thyroid cancers were 285 times pre-Chernobyl levels in Belarus in 1995, especially among children. Belarus's cabinet minister claimed at that time that a quarter of his country's national income was being spent on alleviating the effects of the disaster. As many as 375,000 people in Belarus, Ukraine, and Russia were displaced because of the accident. Contaminated forests spread radioactivity through fires, and seepage from the reactor has polluted waterways as far away as the Black Sea, some 350 miles away.

In April 1996 a fire in the woods near the Chernobyl reactor once more alarmed observers, who feared further danger to the entombed reactor and the release of radioactivity from soils and vegetation in the area. In addition,

FIGURE 5.7

Operational Soviet-designed nuclear power plants, 2000

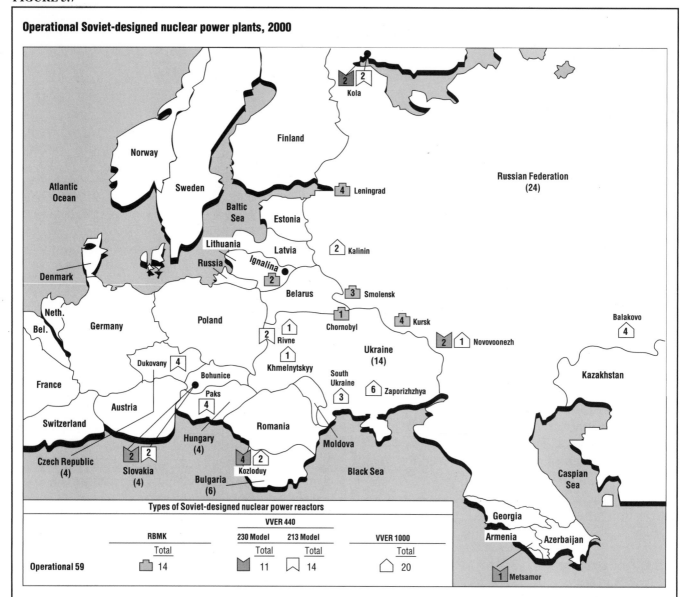

Types of Soviet-designed nuclear power reactors

		VVER 440		
	RBMK	230 Model	213 Model	VVER 1000
	Total	Total	Total	Total
Operational 59	14	11	14	20

Notes:

1. Numbers in parentheses show the total number of reactors in each country, and numbers within symbols show the number of reactors of a specific type at a site.

2. DOE is providing assistance to five other nuclear power reactors in Russia. At one site, Bilibino, four small-scale RBMK reactors produce both steam and electricity. In addition, one fast-neutron reactor is located at Beloyarsk.

SOURCE: "Figure 1: Operational Soviet-Designed Nuclear Power Plants." in *Nuclear Safety: Concerns with the Continuing Operations of Soviet-Designed Nuclear Power Reactors,* U.S. General Accounting Office, Washington, DC, April 2000

the sarcophagus built to contain the damaged reactor is reported to be crumbling. However, the Ukraine, which suffers from 40 percent unemployment and other enormous economic woes, claims it needs the power from the remaining reactor and that completely shutting down Chernobyl would cost too much.

Some high-technology companies have developed robots to clean up hazardous sites such as Chernobyl. Such efforts would, however, produce yet another problem: what to do with the nuclear waste once the Chernobyl sarcophagus is entered and cleaned up.

PROBLEMS WITH SOVIET-BUILT REACTORS. In April 2000 the U.S. General Accounting Office published a report (*Nuclear Safety: Concerns with the Continuing Operations of Soviet-designed Nuclear Power Reactors*) that underscored previous concerns about the danger of Soviet-built reactors. According to U.S. government officials, of the 59 reactors in operation in the former Soviet Union, 25 reactors in the Ukraine, Slovakia, Lithuania, Russia, and Bulgaria pose serious concerns because they fall below Western safety standards and cannot be improved. Although Russian officials earlier claimed that the reactors were "comparable or superior to American designs," they have since admitted that because of economic difficulty throughout the region, funds were not available to repair and staff some facilities. Many of the plants are old and scheduled for retirement, but economic pressures have kept them open. Figure 5.7 shows the locations of Soviet-designed nuclear plants. (Note: In this graphic, Chernobyl is identified by an alternative spelling, Chornobyl.)

A nuclear plant in Metzamor, Armenia, which sits near two major and several minor geologic faults, was closed after an earthquake in 1988. In 1995, despite Western protests, the energy-starved nation restarted the reactor. In turning to the Metzamor reactor for power, Armenia is relying on a Soviet design that U.S. scientists consider among the world's most dangerous. American scientists cite the lack of a dome-shaped containment vessel, which is standard on Western reactors.

Nuclear accidents, wherever they occur, are worldwide problems. Fallout from Chernobyl has been found in northern Europe and in the United States. The international community had pledged nearly $2 billion in nuclear safety assistance as of November 1999. (See Table 5.4.) The United States agreed to send more than $532 million in nuclear safety assistance to countries including Russia, the Ukraine, Bulgaria, Hungary, Lithuania, the Czech Republic, and the Slovak Republic. The U.S. goal for nuclear assistance is to increase the safety of old reactors without lengthening their operational lifetimes, and to find replacement sources of energy so that unsafe reactors can be shut down as quickly as possible.

In Japan

MINAMI. On February 9, 1991, the Minami nuclear power plant in Mihama, Japan (220 miles west of Tokyo), almost suffered a meltdown when a pipe carrying superheated radioactive water broke in the 19-year-old plant. The water leaked, contaminating the water in the steam generator. A reported 20 tons of water were released before the plant shut down. After initially denying any radiation leak, the electric company eventually said that an amount equal to about 8 percent of the plant's annual emissions had been released into the atmosphere. A sister plant was closed the next month because it had the same design flaw that existed at Minami. A storm of protest gave new life to the Japanese antinuclear movement.

The Japanese are in a difficult position concerning nuclear power. Japan has very few natural energy resources and is heavily dependent on Middle Eastern oil. The Japanese government felt nuclear power would provide increasing proportions of its electric energy, but there has been growing opposition from many Japanese people, including survivors of the nuclear bombings of Hiroshima and Nagasaki that ended World War II (1939–45). With several of Japan's biggest nuclear plants out of operation because of safety concerns, the country's demand for

TABLE 5.4

Countries donating and receiving international nuclear safety assistance, as of November 1999

Dollars in millions

Pledged assistance			Received assistance	
Donor	**Amount**		**Recipient**	**Amount**
European Union	$753		Russia	$734
United States	532		Ukraine	629
Germany	168		Bulgaria	132
Japan	139		Lithuania	123
France	91		Regional/Unspecified[a]	58
United Kingdom	45		Slovak Republic	56
Sweden	44		Czech Republic	50
IAEA	43	→	Hungary	42
Canada	26		Armenia	32
Italy	25		NIS Regional[a]	26
Finland	18		Other[b]	35
Norway	14		Kazakhstan	15
Switzerland	13		**Total**	**$1,930**
Belgium	6			
Netherlands	5			
Denmark	4			
Spain	3			
EBRD	1			
Total	**$1,930**			

Notes:
1. Contributions to the Nuclear Safety Account are included in the amounts shown for each donor country.
2. The $532 million listed here as the U.S. contribution differs from the $545 million we identified because of the method used by the G-24 to classify projects and exchange rate variables. In addition, the G-24 data do not include amounts pledged by the United States for the Chornobyl Shelter Implementation Plan.
3. The cumulative contributions of Austria, the World Bank, and the Organization for Economic Cooperation and Development, which total less than $1 million, are not included in the figure.
[a]Nuclear Safety Account funds not yet allocated to specific recipients have been divided equally among these recipients.
[b]Other recipient countries include Azerbaijan, Belarus, Estonia, Georgia, Kyrgyzstan, Latvia, Moldova, Poland, Romania, Slovenia, and Uzbekistan.

SOURCE: "Figure 3: Countries Donating and Receiving International Nuclear Safety Assistance, as of November 1999," in *Nuclear Safety: Concerns with the Continuing Operations of Soviet-Designed Nuclear Power Reactors*, U.S. General Accounting Office, Washington, DC, April 2000

electricity may challenge its generating capacity. Hoping to avoid that, Japan spends three times more than the United States does on nuclear energy research.

A RETREAT FROM NUCLEAR POWER?

In December 1997 the NRC fined Northeast Utilities $2.1 million for a host of violations of federal regulations at three nuclear reactors at the Millstone Nuclear Power Station in Waterford, Connecticut. The fine was nearly double the second-largest fine ever imposed by the NRC, and agency officials investigated criminal prosecution as well. (Previously, the largest fine of $1.25 million was assessed in 1988 against Philadelphia Electric Company's Peach Bottom plant.)

The NRC had previously shut down the three Millstone reactors out of safety concerns, citing more than 50 violations from October 1995 to December 1996. Other plants had undergone reviews, but Millstone was the first instance in which the NRC required the company to hire an independent consulting firm to examine it. Critics of the industry claimed the fine was not severe enough and that revocation of the license would have been a more appropriate penalty—and a message to the rest of the industry. Nevertheless, the fine was another blow to the already troubled nuclear industry. The Millstone I plant was permanently shut down in July 1998.

Deregulation of the electric industry (the removal of governmental regulations and restructuring of the industry to be more competitive) has raised questions about whether expensive nuclear power plants, which must comply with extensive government regulations and control, can compete. Since nuclear energy faces increased competition from cheaper energy sources, regulators worry that nuclear utilities may try to lower expenses by cutting corners on safety measures. The aging of the nation's reactors heightens the need for vigilance. Each year, maintenance problems grow more complex and expensive, raising supervision costs and requirements.

In 1998 Commonwealth Edison permanently closed 2 of its 12 nuclear reactors (in Zion, Illinois) because they were too expensive to operate under industry deregulation. The closing is estimated to cost $515 million. Spent fuel will be stored on site until 2014, when final decommissioning will begin. In addition to Zion, four other Commonwealth Edison plants are on the NRC's watch list. Industry experts predict Commonwealth Edison will be forced to close two of the other sites in order to focus on the remaining reactors. However, other than the Millstone closing in July 1998, as of 2002 no nuclear reactors have been shut down in the United States since the two Zion plants.

THE PRICE-ANDERSON ACT—PAYING FOR A NUCLEAR ACCIDENT

Immediately after the Chernobyl incident in 1987, the Price-Anderson Act (PL95-256, amended), which limited how much a nuclear utility owner would have to pay to cover the costs of a nuclear accident, came up for renewal. At the time, the Price-Anderson Act required nuclear reactor owners to carry $650 million of off-site liability insurance, the amount at which the law limited their liability in the event of a nuclear accident. If damages from an accident exceeded $650 million, all nuclear plants could be charged up to $5 million per reactor to help cover the costs. After the Chernobyl accident, legislation was enacted to increase the liability cap to $7.4 billion.

NEW TECHNOLOGY?

Without new orders for nuclear plants, the U.S. nuclear industry could go out of business in the next few decades; the industry has had no new orders since 1978. Prompted by this grim future, suppliers have rethought plant design and reactor safety. In what government and industry offi-

cials hope will be a new era in nuclear power, they are now proposing smaller, standardized, and more simplified reactors that they claim will be 300 times safer than what current regulations require. These modular plants would take half as long to build, at barely a quarter of the cost of current designs. Designers claim that the plants will be serviced in part by robots and will rely on natural convection for emergency cooling, thus requiring fewer engineered safety features. The smaller plants will also use standardized parts that can be produced in commercial factories, eliminating the need for custom on-site assembly.

NUCLEAR FUSION

On November 1, 1952, a thermonuclear bomb equivalent to 10 million tons of TNT was detonated at Eniwetok Atoll in the Pacific. That moment was both a nightmare and a dream. It showed the potential of even more destruction, as well as the possibility of generating cheap and abundant electricity by fusing hydrogen nuclei together. That dream of safe nuclear energy survives today as scientists experiment with nuclear fusion as a potential major source of power for the twenty-first century.

Fusion is the process by which two atomic particles are joined together under certain conditions to release vast amounts of heat energy. In a fusion reaction, deuterium and tritium atoms fuse, creating helium and energetic neutrons. The neutrons escape through a chamber wall into a surrounding container, which absorbs heat that is used to create steam for generating electricity.

The closest natural example of fusion is found in the Sun, where the temperature at the core is 14 million degrees centigrade. Because of extreme gravitational pressure, nuclei are driven so close together that they fuse and release vast amounts of energy. In a fusion reactor, the temperature must be even higher, around 200 million degrees centigrade, since there is less compression than at the center of the Sun.

Fusion has several advantages over fossil fuels and nuclear fission power. Fusion does not create air pollutants that contribute to acid rain or global warming. Deuterium is available in essentially unlimited supply from seawater, and tritium can be generated on-site as part of the fusion process. In theory, fusion could produce far more energy from the top two inches of Lake Erie than exists in all of the earth's known oil reserves.

From 1982 through 1997 the Princeton Plasma Physics Laboratory operated the most powerful "magnetic confinement" fusion machine in the world: the Tokamak Fusion Test Reactor (TFTR). (Tokamaks are donut-shaped vessels in which an element is compressed and heated to millions of degrees.) The TFTR was one of the world's largest and most successful experimental fusion machines, producing a world-record 10.7 million watts of fusion power, enough to power 3,000 average-sized homes. Additionally, the TFTR produced data that added greatly to understanding of the physics of fusion and helped in the development of the National Spherical Torus Experiment, a new innovative magnetic fusion device constructed by the Princeton Plasma Physics Laboratory in collaboration with the Oak Ridge National Laboratory, Columbia University, and the University of Washington at Seattle.

THE PROBLEM OF NUCLEAR WASTE

Nuclear waste has sometimes been called the Achilles' heel of the nuclear power industry; much of the controversy over nuclear power centers on the lack of a disposal system for the highly radioactive spent fuel that must be regularly removed from operating reactors. As a result, progress on nuclear waste disposal is widely considered a prerequisite for any future growth of nuclear power.

—Congressional Research Service, August 1998

Working in a laboratory in Chicago, Illinois, in 1942, Italian physicist Enrico Fermi assembled enough uranium to cause a nuclear fission reaction. His discovery transformed both warfare and energy production. But his experiment also produced a small packet of radioactive waste materials that will remain dangerous for hundreds of thousands of years. That "first" radioactive waste lies buried under a foot of concrete and two feet of dirt on a hillside in Illinois.

Types of Radioactive Waste

Radioactive waste material is produced at all stages of the nuclear fuel cycle, from the initial mining of the uranium to the final disposal of the spent fuel from the reactor. *Radioactive waste* is a term that encompasses a broad range of material with widely varying characteristics. Some is barely radioactive and safe to handle, while other types are intensely hot in both temperature and radioactivity. Some waste decays to safe levels of radioactivity in a matter of days or weeks, while other types will remain dangerous for thousands of years. The DOE and the NRC define the major types of radioactive waste as follows.

URANIUM MILL TAILINGS. Uranium mill tailings are sandlike wastes produced in uranium refining operations. Although they emit low levels of radiation, their large volumes (10–15 million tons annually) pose a hazard, particularly from radon emissions and groundwater contamination. Since it was not until the early 1970s that the dangers of uranium mill tailings were realized, many miners and residents in the western United States had unsafe exposure to them. Cancer incidences among miners from the pre-1970s period are high. Congress passed the Uranium Mill Tailing Radiation Control Act of 1978 (PL95-604) to clean up abandoned mill sites (primarily at federal expense). Owners of still active mines were financially responsible for their own cleanup.

FIGURE 5.8

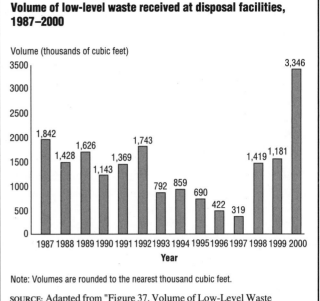

Volume of low-level waste received at disposal facilities, 1987–2000

Note: Volumes are rounded to the nearest thousand cubic feet.

SOURCE: Adapted from "Figure 37. Volume of Low-Level Waste Received at U.S. Disposal Facilities, 1987–2000," in *Information Digest 2001 Edition*, U.S. Nuclear Regulatory Commission, Washington, DC, NUREG 1350, Vol. 13, June 2001

LOW-LEVEL WASTE. Low-level waste, which contains varying lesser levels of radioactivity, includes trash, contaminated clothing, and hardware. In general, low-level waste decays relatively quickly (10–100 years). As of 2000, the DOE was operating six low-level and mixed radioactive waste disposal facilities. Figure 5.8 shows the varying volumes of waste handled at these facilities from 1987 to 2000. A record 3.3 million cubic feet of low-level waste was received in 2000. About 84 percent of it came from industrial sources and 8 percent from utilities. Most of the rest came from the government.

SPENT FUEL. Spent fuel is used reactor fuel that will be classified as waste if not reprocessed to recover the usable uranium and plutonium. In reprocessing, some of the used uranium and plutonium in spent reactor fuel can be removed for use again as nuclear reactor fuel, but even after this process, enough radioactive material remains to make the fuel dangerous for thousands of years. There is currently no long-term, high-level permanent waste disposal repository for spent fuel, but one is being considered at Yucca Mountain in Nevada. In the meantime, spent fuel is being stored on-site at nuclear power plants. Spent fuel is the most radioactive type of civilian nuclear waste.

HIGH-LEVEL WASTE. High-level waste is the by-product of reprocessing plants, containing highly toxic and extremely dangerous fission products. Although most of the uranium and plutonium usually have been removed for reuse, enough long-lived radioactive elements remain to require isolation for 10,000 years or more.

TRANSURANIC (TRU) WASTES. Transuranic (TRU) wastes are 11 human-made radioactive elements with atomic numbers greater than that of uranium (92) and therefore beyond ("trans-") uranium ("-uranic") on the periodic chart of the elements. Their half-lives—the time it takes for half the radioisotopes present in a sample to decay to nonradioactive elements—are thousands of years. They are found in trash produced mainly by nuclear weapons plants and are therefore part of the nuclear waste problem but not directly the concern of nuclear power utilities.

MIXED WASTE. Mixed waste is high-level, low-level, or TRU waste that contains hazardous nonradioactive waste. Such waste poses serious institutional problems, because the radioactive portion is regulated by the DOE or NRC under the Atomic Energy Act, while the Environmental Protection Agency (EPA) regulates the nonradioactive elements under the Resource Conservation and Recovery Act (RCRA; PL 95-510).

Waste Disposal: Necessary but Difficult

Radioactive waste is also a by-product of nuclear weapons plants, hospitals, and scientific research. Disposing of this waste is unquestionably one of the major problems associated with the development of nuclear power. Although federal policy is based on the assumption that nuclear waste can be disposed of safely, new storage and disposal facilities for all types of radioactive waste have frequently been delayed or blocked by concerns about safety, health, and the environment.

The highly toxic wastes must be isolated from the environment until the radioactivity decays to a safe level. In the case of plutonium, for example, the half-life is 26,000 years. At that rate, it will take at least 100,000 years before radioactive plutonium is no longer dangerous. Any facilities to store such materials must last at least that long.

Nuclear Waste Legislation

A major step toward shifting the responsibility for disposal of high-level nuclear wastes from the nuclear power industry to the federal government was taken in 1982, when Congress passed the Nuclear Waste Policy Act (PL 97-425). It provided the first national, comprehensive policy and detailed timetable for the management and disposal of high-level nuclear waste and authorized construction of the first nuclear waste repository. A 1987 amendment to the Nuclear Waste Policy Act directed investigation of the Yucca Mountain site in Nevada as a potential repository. On July 9, 2002, the U.S. Senate joined the U.S. House of Representatives in approving the Yucca Mountain site. The next step in developing the facility is for independent experts at the NRC to review the 24 years of scientific study of Yucca Mountain and to consider the site for a license.

Federal Nuclear Waste Repositories: The Promise of Yucca Mountain

The U.S. government is focusing on two locations as eventual long-term nuclear waste repositories: the Waste Isolation Pilot Plant (WIPP) in southeastern New Mexico for transuranic waste (which is primarily derived from military activities and not from nuclear power plants) and Nevada's Yucca Mountain for civilian waste.

Figure 5.9 shows the location of the proposed radioactive waste disposal facility at Yucca Mountain. The NRC must be convinced by the DOE that the Yucca Mountain facility would meet EPA standards for confining radioactivity. The NRC will consider the site for a license after its review of the scientific data. Repository construction cannot begin until the NRC grants a construction authorization. The earliest date that the department could begin building the repository is 2005. It would take another 5 years to complete construction and begin operations, 12 years later than originally expected.

The proposed repository would look like a large mining complex. It would have facilities on the surface for handling and packaging nuclear waste and a large mine about 1,000 feet underground. Here, plans call for the waste to be placed in sealed metal canisters placed vertically in the floor of underground tunnels. Aboveground facilities would cover approximately 400 acres and be surrounded by a three-mile buffer zone. Underground, about 1,400 acres would be mined, consisting of tunnels leading to the areas where waste containers would be placed and service areas near the shafts and ramps that provide access from the surface. This type of "deep geologic" disposal is widely considered by governments, scientists, and engineers to be the best option for isolating highly radioactive waste.

The repository would be designed to contain radioactive material by using layers of human-made and natural barriers. Regulations require that a repository isolate waste until the radiation decays to a level that is about the same as that from a natural underground uranium deposit. This decay time is estimated at about 10,000 years.

After the repository has been filled to capacity, regulations require the DOE to keep the facility open and to monitor it for at least 50 years from the fill date. This will allow experts to monitor conditions inside the repository and retrieve spent fuel if necessary. Eventually, the repository shafts will be filled with rock and earth and sealed. At the ground level, facilities will be removed, and as much as possible, the DOE will take steps to return the site to its original condition.

Scientists assume that over thousands of years, some of the human-made barriers in a repository will break down. Once that happens, natural barriers will be counted on to stop or slow the movement of radiation particles. The most

FIGURE 5.9

Yucca Mountain waste site

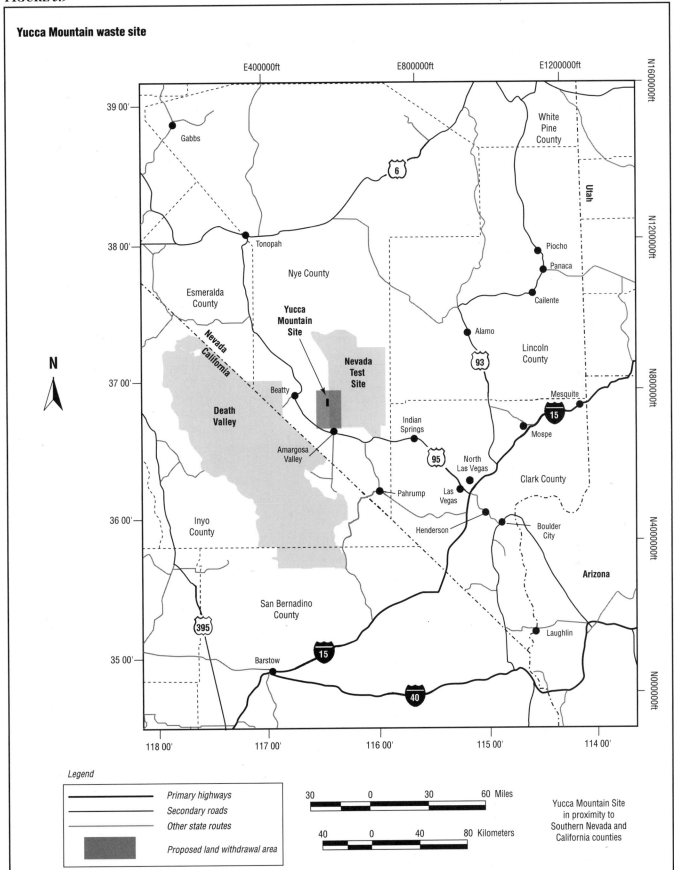

Legend

━━━━━━━ Primary highways
─────── Secondary roads
─────── Other state routes
▨ Proposed land withdrawal area

30   0   30   60 Miles

40   0   40   80 Kilometers

Yucca Mountain Site
in proximity to
Southern Nevada and
California counties

SOURCE: "Figure 1-5. Map Showing the Location of Yucca Mountain in Relation to Major Highways; Surrounding Counties, Cities, and Towns in Nevada and California; the Nevada Test Site; and Death Valley National Park," in *Yucca Mountain Science and Engineering Report: Technical Information Supporting Site Recommendation Consideration Revision 1*, U.S. Department of Energy, Washington, DC, February 2002

TABLE 5.5

Total commercial spent nuclear fuel discharges, 1968–98

Reactor Type	Stored at Reactor Sites	Number of Assemblies Stored at Away-from-Reactor Facilities	Total
Boiling-Water Reactor	73,538	2,957	76,495
Pressurized-Water Reactor	56,778	491	57,269
High-Temperature Gas Cooled Reactor	1,464	744	2,208
Total	**131,780**	**4,192**	**135,972**
	Metric Tonnes of Uranium (MTU)		
Boiling-Water Reactor	13,230.3	554.0	13,784.2
Pressurized-Water Reactor	24,412.7	192.6	24,605.4
High-Temperature Gas Cooled Reactor	15.4	8.8	24.2
Total	**37,658.3**	**755.4**	**38,413.7**

MTU = Metric tonnes of uranium.
Notes: A number of assemblies discharged prior to 1972, which were reprocessed, are not included in this table (no data is available for assemblies reprocessed before 1972). Totals may not equal sum of components because of independent rounding.

SOURCE: "Table 1. Total U.S. Commercial Spent Nuclear Fuel Discharges, 1968–1998," in *Detailed United States Spent Nuclear Fuel Data as of December 31, 1998,* U.S. Department of Energy, Energy Information Administration, Washington, DC, [Online] http://www.eia.doe.gov/cneaf/nuclear/spent_fuel/ussnfdata.html [accessed September 18, 2002]

likely way for particles to reach humans and the environment would be through water, which is why the low water tables at Yucca Mountain are so crucial. Yucca Mountain also has certain chemical properties that act as another barrier to the movement of radioactive particles. Minerals in the rock called zeolites would stick to the particles and slow their movement throughout the environment.

A Waste Crisis in the Nuclear Power Industry

The long delay in providing disposal sites for nuclear wastes, coupled with the accelerated pace at which nuclear plants are being retired, has created a crisis in the industry. Several aging plants are being maintained at a cost of $20 million a year for each reactor simply because there is no place to send the waste once the plants are decommissioned. Table 5.5 shows that as of 1998 more than 38,000 metric tons of nuclear uranium waste were sitting in spent fuel pools at the 109 operating and 16 closed nuclear energy plants around the country, and this amount is growing. When a nuclear plant shuts down, the nuclear waste and the radioactive equipment stay on the premises because there is no place to put them. As a result, every nuclear power plant in the United States has become a temporary nuclear waste disposal site. Plants that close become mausoleums, largely untouched while they wait to be decommissioned or dismantled when a repository opens.

According to the NRC, almost all nuclear plants will soon reach their capacity for storing their spent nuclear waste because they already have a backlog of waste. Additionally, many are vulnerable to shutdown because production costs are higher than the projected market prices of electricity. Moreover, many nuclear plants do not operate for the 40 years for which they are licensed. Of the 16 reactors that have been shut down, 1 was shut down at 34 years, 6 at 24 to 28 years, 6 at 10 to 18 years, and 3 at under 10 years.

CHAPTER 6
RENEWABLE ENERGY

RENEWABLE ENERGY DEFINED

Imagine energy sources that use no oil, produce no pollution, cannot be affected by political events and cartels, create no radioactive waste, and yet are economical. Although it sounds impossible, some experts claim that technological advances could make wide use of renewable energy sources possible within a few decades. Renewable energy is energy that is naturally regenerated and is, therefore, virtually unlimited. Sources include the Sun (solar), wind, water (hydropower), vegetation (biomass), and the heat of the earth (geothermal).

Solar energy, wind energy, hydropower, and geothermal power are all renewable, inexpensive, and clean sources of energy. Each of these alternative energy sources has advantages and disadvantages, and many observers hope that one or more of them may eventually provide a substantially better energy source than conventional fossil fuels. As the United States and the rest of the world continue to expand their energy needs, putting a strain on the environment and nonrenewable resources, alternative sources of energy continue to be explored.

A HISTORICAL PERSPECTIVE

Before the eighteenth century, most energy came from renewable sources. People burned wood for heat, used sails to harness the wind and propel boats, and installed water wheels on streams to grind grain. The large-scale shift to nonrenewable energy sources began in the 1700s with the Industrial Revolution, a period marked by the rise of factories first in Europe and then in North America. As demand for energy grew, coal replaced wood as the main fuel. Coal was the most efficient fuel for the steam engine, one of the most important inventions of the Industrial Revolution.

Until the early 1970s most Americans were unconcerned about the sources of the nation's energy. Supplies of coal and oil, which together provided more than 90 percent of U.S. energy, were believed to be plentiful. The decades preceding the 1970s were characterized by cheap gasoline and little public discussion of energy conservation.

That carefree approach to energy consumption ended in the 1970s. A fuel oil crisis made Americans more aware of the importance of developing alternative sources of energy to supplement and perhaps eventually even replace fossil fuels. In major cities throughout the United States, gasoline rationing became commonplace, lower heat settings for offices and living quarters were encouraged, and people waited in line to fill their gas tanks. In a country where mobility and personal transportation were highly valued, the oil crisis was a shocking reality for many Americans. As a result, the administration of President Jimmy Carter (1977–81) encouraged federal funding for research into alternative energy sources.

In 1978 the U.S. Congress passed the Public Utilities Regulatory Policies Act (PURPA; PL 95-617), which was designed to help the struggling alternative energy industry. The act exempted small alternative producers from state and federal utility regulations and required existing local utilities to buy electricity from them. The renewable energy industries responded by growing rapidly, gaining experience, improving technologies, and lowering costs. This act was the single most important factor in the development of the commercial renewable energy market.

In the 1980s President Ronald Reagan decided that private-sector financing for the short-term development of alternative energy sources was better than public-sector financing. As a result, he proposed the reduction or elimination of federal expenditures for alternative energy sources. Although funds were severely cut, the U.S. Department of Energy (DOE) continued to support some research and development to explore alternate sources of energy. The Clinton administration reemphasized the importance of renewable energy and increased funding in several areas. The George W. Bush administration believed that renewable and alternative fuels offer hope for the future but also deemed that renewables contribute only a small portion of America's energy needs today. The Bush administration supported funding for research and development in renewable technologies and tax credits for the purchase of hybrid and fuel cell cars.

DOMESTIC RENEWABLE ENERGY USAGE

Renewable energy contributes only a small portion of the nation's energy supply. In 2000 the United States consumed an estimated 6.8 quadrillion Btu of renewable energy, about 7 percent of the nation's total energy consumption. (See Table 6.1 and Figure 6.1.) Biomass sources (wood and waste) contributed 3.3 quadrillion Btu, while hydropower provided 3.1 quadrillion Btu. Together, biomass and hydropower provided 92 percent of renewable energy in 2000, or around 6 percent of all energy. Geothermal energy was the third largest source, with about 0.3 quadrillion Btu. Solar power contributed 0.07 quadrillion Btu, and wind provided 0.05 quadrillion Btu. Electric utilities and the industrial sector were the biggest consumers of renewable energy in 2000.

BIOMASS ENERGY

Biomass refers to organic material such as plant and animal waste, wood, seaweed and algae, and garbage. The use of biomass is not without environmental problems. Deforestation can occur from widespread wood use if forests are clear-cut, resulting in the possibility of soil erosion and mudslides. Burning wood, like burning fossil fuels, also pollutes the environment. Biomass can be burned directly or converted to biofuel by thermochemical conversion and biochemical conversion.

Direct Burning

Direct combustion is the easiest and most commonly used method of using biomass as fuel. Materials such as dry wood or agricultural wastes are chopped and burned to produce steam, electricity, or heat for industries, utilities, and homes. Industrial-size wood boilers are operating throughout the country, and the DOE maintains that many more will be built during the next decade. The burning of agricultural wastes is also becoming more widespread. In Florida, sugarcane producers use the residue from the cane to generate much of their energy.

Wood burning in stoves and fireplaces is another example of direct burning of biomass for energy, in this case heat. In the United States, residential use of wood as fuel generated 433 trillion Btu in 2000. In comparison, the generation of Btu from burning wood in the home in the 1980s and early-to-mid- 1990s was 800–900 trillion Btu.

Thermochemical Conversion

Thermochemical conversion involves heating biomass in an oxygen-free or low-oxygen atmosphere, transforming the material into simpler substances that can be

FIGURE 6.1

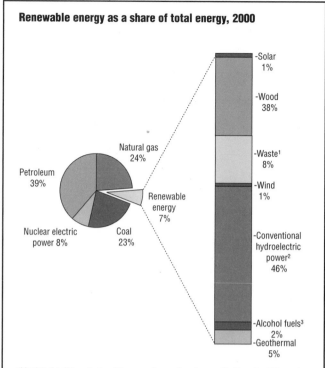

Renewable energy as a share of total energy, 2000

Petroleum 39%
Natural gas 24%
Nuclear electric power 8%
Coal 23%
Renewable energy 7%

Solar 1%
Wood 38%
Waste[1] 8%
Wind 1%
Conventional hydroelectric power[2] 46%
Alcohol fuels[3] 2%
Geothermal 5%

[1]Municipal solid waste, landfill gas, methane, digester gas, liquid acetonitrile waste, tall oil, waste alcohol, medical waste, paper pellets, sludge waste, solid byproducts, tires, agricultural byproducts, closed loop biomass, fish oil, and straw.
[2]Includes electricity net imports derived from hydroelectric power. Before 1989, includes net imports derived from all resources.
[3]Includes ethanol blended into motor gasoline.

SOURCE: Adapted from "Figure 10.1. Renewable Energy Consumption by Source," in *Annual Energy Review 2000,* U.S. Department of Energy, Energy Information Administration, Washington, DC, 2001.

used as fuels. Products such as charcoal and methanol are produced this way.

Biochemical Conversion

Biochemical conversion uses enzymes, fungi, or other microorganisms to convert high-moisture biomass into either liquid or gaseous fuels. Bacteria convert manure, agricultural wastes, paper, and algae into methane, which is used as fuel. Sewage treatment plants have used anaerobic (oxygen-free) digestion for many years to generate methane gas. Small-scale digesters have been used on farms, primarily in Europe and Asia, for hundreds of years. In 2002 it was estimated that China had 10 million biogas pits (a biomass-based technology).

Another type of biochemical conversion process, fermentation, uses yeast to decompose carbohydrates, yielding ethyl alcohol (ethanol) and carbon dioxide. Sugar crops, grains (corn, in particular), potatoes, and other starchy crops commonly supply the sugar for ethanol production.

BIOMASS CONVERSION PRODUCTS

Ethanol and Methanol

Ethanol (ethyl alcohol) is a colorless, nearly odorless, flammable liquid derived from fermenting plant material

TABLE 6.1

Energy consumption by source, 1949–2000

(quadrillion btu)

Year	Fossil fuels						Nuclear electric power	Hydro-electric pumped storage[5]	Renewable energy[1]						Total[7]
	Coal	Coal coke net imports[2]	Natural gas[2]	Petroleum	Electricity net imports[4]	Total			Conventional hydroelectric power[6]	Wood, waste, alcohol[8]	Geothermal[8]	Solar	Wind	Total	
1949	11.981	-0.007	5.145	11.883	(9)	29.002	0	(9)	1.449	1.549	NA	NA	NA	2.998	32.000
1950	12.347	0.001	5.968	13.315	(9)	31.632	0	(9)	1.440	1.562	NA	NA	NA	3.003	34.635
1951	12.553	-0.021	7.049	14.428	(9)	34.008	0	(9)	1.454	1.535	NA	NA	NA	2.988	36.996
1952	11.306	-0.012	7.550	14.956	(9)	33.800	0	(9)	1.496	1.474	NA	NA	NA	2.970	36.770
1953	11.373	-0.009	7.907	15.556	(9)	34.826	0	(9)	1.439	1.419	NA	NA	NA	2.857	37.684
1954	9.715	-0.007	8.330	15.839	(9)	33.877	0	(9)	1.388	1.394	NA	NA	NA	2.783	36.660
1955	11.167	-0.010	8.998	17.255	(9)	37.410	0	(9)	1.407	1.424	NA	NA	NA	2.832	40.242
1956	11.350	-0.013	9.614	17.937	(9)	38.888	0	(9)	1.487	1.416	NA	NA	NA	2.903	41.791
1957	10.821	-0.017	10.191	17.932	(9)	38.926	(s)	(9)	1.557	1.334	NA	NA	NA	2.890	41.816
1958	9.533	-0.007	10.663	18.527	(9)	38.717	0.002	(9)	1.629	1.323	NA	NA	NA	2.952	41.670
1959	9.518	-0.008	11.717	19.323	(9)	40.550	0.002	(9)	1.587	1.353	NA	NA	NA	2.940	43.493
1960	9.838	-0.006	12.385	19.919	(9)	42.137	0.006	(9)	1.657	1.320	0.001	NA	NA	2.977	45.120
1961	9.623	-0.008	12.926	20.216	(9)	42.758	0.020	(9)	1.680	1.295	0.002	NA	NA	2.977	45.755
1962	9.906	-0.006	13.731	21.049	(9)	44.681	0.026	(9)	1.822	1.300	0.002	NA	NA	3.124	47.832
1963	10.413	-0.007	14.403	21.701	(9)	46.509	0.038	(9)	1.772	1.323	0.004	NA	NA	3.099	49.647
1964	10.964	-0.010	15.288	22.301	(9)	48.543	0.040	(9)	1.907	1.337	0.005	NA	NA	3.248	51.831
1965	11.581	-0.018	15.769	23.246	(9)	50.577	0.043	(9)	2.058	1.335	0.004	NA	NA	3.397	54.016
1966	12.143	-0.025	16.995	24.401	(9)	53.514	0.064	(9)	2.073	1.369	0.004	NA	NA	3.446	57.024
1967	11.914	-0.015	17.945	25.284	(9)	55.127	0.088	(9)	2.344	1.340	0.007	NA	NA	3.691	58.906
1968	12.331	-0.017	19.210	26.979	(9)	58.502	0.142	(9)	2.342	1.419	0.009	NA	NA	3.771	62.415
1969	12.382	-0.036	20.678	28.338	(9)	61.362	0.154	(9)	2.659	1.440	0.013	NA	NA	4.113	65.628
1970	12.265	-0.058	21.795	29.521	(9)	63.522	0.239	(9)	2.654	R1.431	0.011	NA	NA	R4.096	R67.858
1971	11.598	-0.033	22.469	30.561	(9)	64.596	0.413	(9)	2.861	R1.432	0.012	NA	NA	R4.305	R69.314
1972	12.077	-0.026	22.698	32.947	(9)	67.696	0.584	(9)	2.944	R1.503	0.031	NA	NA	R4.478	R72.758
1973	12.971	-0.007	22.512	34.840	(9)	70.316	0.910	(9)	3.010	R1.529	0.043	NA	NA	R4.581	R75.808
1974	12.663	0.056	21.732	33.455	(9)	67.906	1.272	(9)	3.309	R1.540	0.053	NA	NA	R4.902	R74.080
1975	12.663	0.014	19.948	32.731	(9)	65.355	1.900	(9)	3.219	R1.499	0.070	NA	NA	R4.788	R72.042
1976	13.584	(s)	20.345	35.175	(9)	69.104	2.111	(9)	3.066	R1.713	0.078	NA	NA	R4.857	R76.072
1977	13.922	0.015	19.931	37.122	(9)	70.989	2.702	(9)	2.515	R1.838	0.077	NA	NA	R4.431	R78.122
1978	13.766	0.125	20.000	37.965	(9)	71.856	3.024	(9)	3.141	R2.038	0.064	NA	NA	R5.243	R80.123
1979	15.040	0.063	20.666	37.123	(9)	72.892	2.776	(9)	3.141	R2.152	0.084	NA	NA	R5.377	R81.044
1980	15.423	-0.035	20.394	34.202	(9)	69.984	2.739	(9)	3.118	R2.485	0.110	NA	NA	R5.712	R78.435
1981	15.908	-0.016	19.928	31.931	(9)	67.750	3.008	(9)	3.105	2.590	0.123	NA	NA	5.818	76.569
1982	15.322	-0.022	18.505	30.232	(9)	64.037	3.131	(9)	3.572	2.615	0.105	NA	NA	6.292	73.441
1983	15.894	-0.016	17.357	30.054	(9)	63.290	3.203	(9)	3.899	2.831	0.129	NA	(s)	6.860	73.317

TABLE 6.1

Energy consumption by source, 1949–2000 [CONTINUED]

(quadrillion btu)

Year	Fossil fuels						Nuclear electric power	Hydro-electric pumped storage⁵	Renewable energy¹						Total⁷
	Coal	Coal coke net imports	Natural gas²	Petroleum	Electricity net imports⁴	Total			Conventional hydroelectric power⁶	Wood, waste, alcohol⁷	Geothermal⁸	Solar	Wind	Total	
1984	17.071	-0.011	18.507	31.051	(9)	66.617	3.553	(9)	3.800	2.880	0.165	(s)	(s)	6.845	76.972
1985	17.478	-0.013	17.834	30.922	(9)	66.221	4.149	(9)	3.398	R2.864	0.198	(s)	(s)	R6.460	R76.778
1986	17.260	-0.017	16.708	32.196	(9)	66.148	4.471	(9)	3.446	R2.841	0.219	(s)	(s)	R6.507	77.065
1987	18.008	0.009	17.744	32.865	(9)	68.626	4.906	(9)	3.117	R2.823	0.229	(s)	(s)	R6.170	79.633
1988	18.846	0.040	18.552	34.222	(9)	71.660	5.661	(9)	2.662	2.937	0.217	(s)	(s)	5.817	R83.068
1989	R19.043	0.030	19.384	34.211	R-0.050	R72.618	5.677	(9)	R3.014	R3.060	R0.334	0.059	0.024	R6.492	R84.716
1990	R19.253	0.005	19.296	33.553	R-0.080	R72.027	6.162	-0.036	R3.146	R2.660	R0.355	0.063	0.032	R6.254	R84.344
1991	R18.998	0.010	19.606	32.845	R0.059	R71.519	6.580	-0.047	R3.159	R2.700	R0.363	0.066	0.032	R6.320	R84.298
1992	R19.152	0.035	20.131	33.527	R0.053	R72.897	6.608	-0.043	R2.818	R2.845	R0.374	0.067	0.032	R6.134	R85.513
1993	R19.763	0.027	20.827	33.841	R0.050	R74.508	6.520	-0.042	R3.119	R2.803	R0.387	0.071	0.030	R6.410	R87.300
1994	R19.933	0.058	21.288	34.670	R0.140	R76.089	6.838	-0.035	R2.993	R2.938	R0.391	0.072	0.036	R6.429	R89.213
1995	R20.025	0.061	22.163	34.553	R0.121	R76.924	7.177	-0.028	R3.481	R3.066	R0.333	0.073	0.033	R6.986	R90.943
1996	R20.957	0.023	22.559	35.757	R0.109	R79.406	7.168	-0.032	R3.892	R3.126	R0.346	0.075	0.035	R7.473	R93.931
1997	R21.464	0.046	22.530	36.266	R0.109	R80.415	6.678	-0.042	R3.961	R3.004	R0.322	0.074	R0.033	R7.395	R94.340
1998	21.667	0.067	21.921	36.934	R0.048	80.637	7.157	-0.046	R3.569	R2.976	R0.328	0.074	0.031	R6.977	R94.608
1999	R21.693	0.058	R22.289	R37.960	R0.092	R82.090	R7.736	R-0.065	R3.512	R3.221	R0.373	R0.073	R0.046	R7.226	R96.866
2000ᴾ	22.407	0.065	23.325	37.964	0.102	83.863	8.009	-0.058	3.107	3.275	0.319	0.070	0.051	6.823	98.498

¹End-use consumption, electric utility and nonutility electricity net generation, and net imports of electricity from renewable energy.
²Includes supplemental gaseous fuels.
³Petroleum products supplied, including natural gas plant liquids and crude oil burned as fuel.
⁴Electricity net imports from fossil fuels. May include some nuclear-generated electricity.
⁵Pumped storage facility production minus energy used for pumping.
⁶Through 1988, includes all electricity net imports. From 1989, includes only electricity net imports derived from hydroelectric power.
⁷Alcohol (ethanol blended into motor gasoline) is included in both "Petroleum" and "Alcohol," but is counted only once in total energy consumption.
⁸From 1989, includes electricity imports from Mexico that are derived from geothermal energy.
⁹Included in conventional hydroelectric power.
R=Revised. P=Preliminary. (s)=Less than 0.0005 and greater than -0.0005 quadrillion Btu. NA=Not available.
Totals may not equal sum of components due to independent rounding.

SOURCE: "Table 1.3. Energy Consumption by Source, 1949–2000," in *Annual Energy Review 2000*, U.S. Department of Energy, Energy Information Administration, Washington, DC, 2001

FIGURE 6.2

The carbon cycle

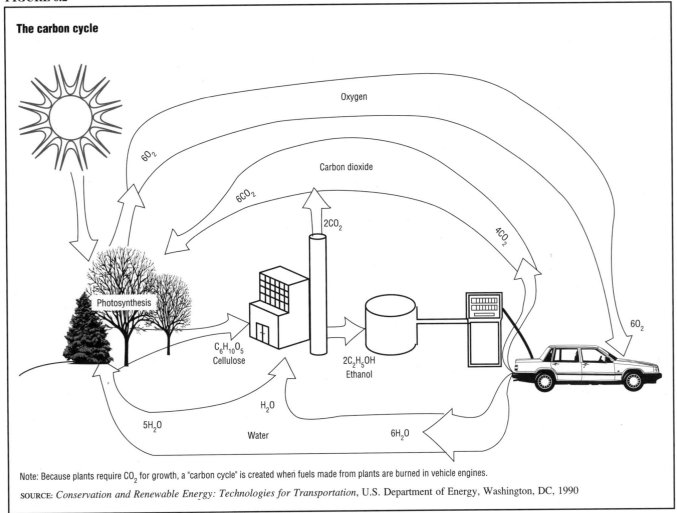

Note: Because plants require CO_2 for growth, a "carbon cycle" is created when fuels made from plants are burned in vehicle engines.

SOURCE: *Conservation and Renewable Energy: Technologies for Transportation*, U.S. Department of Energy, Washington, DC, 1990

that contains carbohydrates in the form of sugar. Most of the ethanol manufactured for use as fuel is derived from corn, wood, and sugar. A mixture of 10 percent ethanol and 90 percent gasoline, called gasohol, is commonly found in the United States. Automobiles can run on gasohol and can be built to run directly on ethanol or on any mixture of ethanol and gasoline. However, ethanol is difficult and expensive to produce in bulk. The development of ethanol as a fuel source may depend more upon the political support of legislators from farming states than upon any real economic benefits.

Some scientists believe ethanol made from wood, sawdust, corncobs, or rice hulls could liberate the alcohol fuel industry from its dependence on food crops, such as corn and sugarcane. Worldwide, enough corncobs and rice hulls are left over from annual crop production to produce more than 40 billion gallons of ethanol.

Advocates of wood-derived ethanol believe that it could eventually be a sustainable liquid fuel industry that does not rely on pollution-generating fossil fuels. For instance, if new trees were planted to replace those that were cut for fuel, they would be available for later har-

vesting while at the same time alleviating global warming with their carbon dioxide–processing function. However, other scientists warn that an increased demand for wood for transportation fuels might accelerate the destruction of old-growth forests and endanger ecosystems.

Methanol (methyl alcohol) fuels have also been tested successfully. Using methanol instead of diesel fuel virtually eliminates sulfur emissions and reduces other environmental pollutants usually emitted from trucks and buses. Producing methanol from biofuels, however, is costly.

Burning biofuels in vehicle engines creates a "carbon cycle" in which the earth's vegetation can, in turn, make use of the products of automobile combustion. (See Figure 6.2.) Automobile combustion generated from fossil fuels, however, contains pollutants.

Municipal Waste Recovery

Each year millions of tons of garbage are buried in landfills and city dumps. This method of disposal is becoming increasingly costly as many landfills across the nation near capacity. Many communities discovered that they could solve both problems, cost and capacity, at once by constructing waste-to-energy plants. Not only is the garbage burned and reduced in volume by 90 percent, but energy in the form of steam or electricity is generated in a cost-effective way, and the potential energy benefit is significant. Use of municipal waste as fuel has increased steadily since the 1980s. According to EIA figures, municipal waste (including landfill gas) generated 88 trillion Btu of energy in 1981, which grew to 551 trillion Btu by 2001.

The two most common waste-to-energy plant designs are the mass burn (also called direct combustion) system and the refuse derived fuel (RDF) system.

MASS BURN SYSTEMS. Most waste-to-energy plants in the United States use the mass burn system. This system's advantage is that the waste does not have to be sorted or prepared before burning, except for removing obviously noncombustible, oversized objects. The mass burn eliminates expensive sorting, shredding, and transportation machinery that may be prone to break down.

In mass burn systems, waste is carried to the plant in trash trucks and dropped into a storage pit. Large overhead cranes lift the garbage into a furnace feed hopper that controls the amount and rate of waste that is fed into the furnace. Next, the garbage is moved through a combustion zone so that it burns to the greatest extent possible. The burning garbage produces heat, and that heat is used to produce steam. The steam can be used directly for industrial needs or can be sent through a turbine to power a generator to produce electricity.

REFUSE DERIVED FUEL (RDF). RDF systems process waste to remove noncombustible objects and to create homogeneous and uniformly sized fuel. Large items such as bedsprings, dangerous materials, and flammable liquids are removed by hand. The trash is then shredded and carried to a screen to remove glass, rocks, and other material that cannot be burned. The remaining material is usually sifted a second time with an air separator to yield fluff. The fluff is sent to storage bins before being burned, or it can be compressed into pellets or briquettes for long-term storage. This fuel can be used as an energy source by itself in a variety of systems, or it can be used with other fuels, such as coal or wood.

PERFORMANCE OF WASTE-TO-ENERGY SYSTEMS. Most waste-to-energy systems can produce two to four pounds of steam for every pound of garbage burned. A 1,000-ton-per-day mass burn system will burn an average of 310,250 tons of trash each year and will recover 2 trillion Btu of energy. In addition, the plant will emit 96,000 tons of ash (32 percent of waste input) for landfill disposal. An RDF plant produces less ash but sends almost the same amount of waste to the landfill because of the noncombustibles that accumulate in the separation process before burning.

DISADVANTAGES OF WASTE-TO-ENERGY PLANTS. The major problem with increasing the use of municipal waste-to-energy plants is their effect on the environment. The emission of particles into the air is partially controlled by electrostatic precipitators, and many gases can be eliminated by proper combustion techniques. There is concern, however, over the amounts of dioxin (a very dangerous air pollutant) and other toxins that are often emitted from these plants. Noise from trucks, fans, and processing equipment at these plants can also be unpleasant for nearby residents.

Landfill Gas Recovery

Landfills contain a large amount of biodegradable matter that is compacted and covered with soil. Bacteria called methanogens thrive in this oxygen-depleted environment. They metabolize the biodegradable matter in the landfill, producing methane gas and carbon dioxide as byproducts. In the past, as landfills aged, these gases built up and leaked out. This gas leakage prompted some communities to drill holes in landfills and burn off the methane to prevent dangerously large amounts from igniting.

The energy crisis of the 1970s made landfill methane gas an energy resource too valuable to waste, and efforts were made to find an inexpensive way to tap the gas. The first landfill gas-recovery site was finished in 1975 at the Palos Verdes Landfill in Rolling Hills Estates, California.

In a typical operation, garbage is allowed to decompose for several months. When a sufficient amount of methane gas has developed, it is piped out to a generating plant, where it is turned into electricity. In its purest form, methane gas is equivalent to natural gas and can be used in exactly the same way. Depending on the extraction rates, most sites can produce gas for about 20 years. The advantages of tapping gas from a landfill go beyond the energy provided by the methane, as extraction reduces landfill odors and the chances of explosions.

HYDROPOWER

Hydropower, the energy that comes from the natural flow of water, is the world's largest renewable energy source. The energy of falling water or flowing water is converted into mechanical energy and then to electrical energy. In the past, flowing water turned waterwheels to grind grain or turn saws, but today flowing water is used to turn modern turbines. Hydropower is a renewable, nonpolluting, and reliable energy source.

Advantages and Disadvantages

At present, hydropower is the only means of storing large quantities of electrical energy for almost instant use. This is done by holding water in a large reservoir behind a dam, with a hydroelectric power plant below. The dam creates a height from which water flows. The fast-moving water pushes the turbine blades that turn the rotor part of the electric generator. Whenever power is needed at peak times, the valves are opened, and in a short amount of time, turbine generators produce power.

Nearly all the best sites for large hydropower plants are already being put to use in the United States. Small hydropower plants are expensive to build but eventually become cost-efficient because of their low operating costs. One of the disadvantages of small hydropower generators is their reliance on rain and melting snow to fill reservoirs, because some years bring drought conditions. Additionally, U.S. environmental groups strongly protest the construction of new dams in America. Ecologists express concern that dams ruin streams, dry up waterfalls, and interfere with marine life habitats.

New Directions in Hydropower Energy

The United States and Europe have developed a major proportion of their hydroelectric potential. Large-scale hydropower development has slowed considerably in the United States. The last federally funded hydropower dam constructed in the United States was the Corps of Engineers' Richard B. Russell Dam and Lake, which is located on the Savannah River and borders South Carolina and Georgia. The project was authorized in 1966 and completed in 1986. However, expansion and efficiency improvements at existing dams still offer significant potential for additional hydropower capacity and energy. Until recently in the United States, dams were usually funded entirely with federal monies. Since 1986, however, half of the cost of any new dam proposed in the United States must be put up by local governments. Hydropower's contribution to U.S. energy generation should remain relatively constant, although existing sites can become more efficient as new generators are added. Any new major supplies of hydroelectric power for the United States will likely come from Canada.

Most of the new development in hydropower is occurring in the Third World, as developing nations see it as an effective method of supplying power to growing populations. Most of these hydropower-development programs are massive public-works projects requiring huge amounts of money, which is mostly borrowed from the developed world. Third World leaders believe that hydroelectric dams are worth the cost and potential environmental threats because they bring cheap electric power to their citizenry.

GEOTHERMAL ENERGY

Since ancient times, humans have exploited the earth's natural hot water sources. Although bubbling hot springs became public baths as early as ancient Rome, using hot water and underground steam to produce power is a relatively recent development. Electricity was first generated from natural steam in Italy in 1904. The world's first steam power plant was built in 1958 in a volcanic region of New Zealand. A field of 28 geothermal power plants covering 30 square miles in northern California was completed in 1960.

What Is Geothermal Energy?

Geothermal energy is the natural, internal heat of the earth trapped in rock formations deep underground. Only a fraction of this vast storehouse of energy can be extracted, usually through large fractures in the earth's crust. Hot springs, geysers, and fumaroles (holes in or near volcanoes from which vapor escapes) are the most easily exploitable sources of geothermal energy. (See Figure 6.3.) Geothermal reservoirs provide hot water or steam that can be used for heating buildings, processing food, and generating electricity.

To produce power from a geothermal energy source, pressurized steam or hot water is extracted from the earth and directed toward turbines. The electricity produced by turbines is then fed into a utility grid and distributed to residential and commercial customers. According to the Energy Information Administration's *Annual Energy Review 2000* (2001), electricity production accounted for 96 percent of the world's geothermal energy commercial use.

Types of Geothermal Energy

Like most natural energy sources, geothermal energy is usable only when it is concentrated in one spot in what is called a "thermal reservoir." The four basic categories of thermal reservoirs are hydrothermal reservoirs, dry rock, geopressurized reservoirs, and magma resources. Most of the known reservoirs for geothermal power in the United States are located west of the Mississippi River. (See Figure 6.4.)

HYDROTHERMAL RESERVOIRS. Hydrothermal reservoirs consist of a heat source covered by a permeable formation through which water circulates. Steam is produced when hot water boils underground and some of the steam escapes to the surface under pressure. Once at the surface, impurities and tiny rock particles are removed, and the steam is piped directly to the electrical generating station. These systems are the cheapest and simplest form of geothermal energy. The Geysers, 90 miles north of San Francisco, California, are the most famous example of this type. As of 2002 the Geysers produced enough electricity to meet the needs of about 1.3 million people. The Geysers Geothermal Field is the world's largest source of geothermal power.

DRY ROCK. Dry rock formations are the most common geothermal source, especially in the West. To tap this source of energy, water is injected into hot rock formations that have been fractured to produce steam or water for collection.

FIGURE 6.3

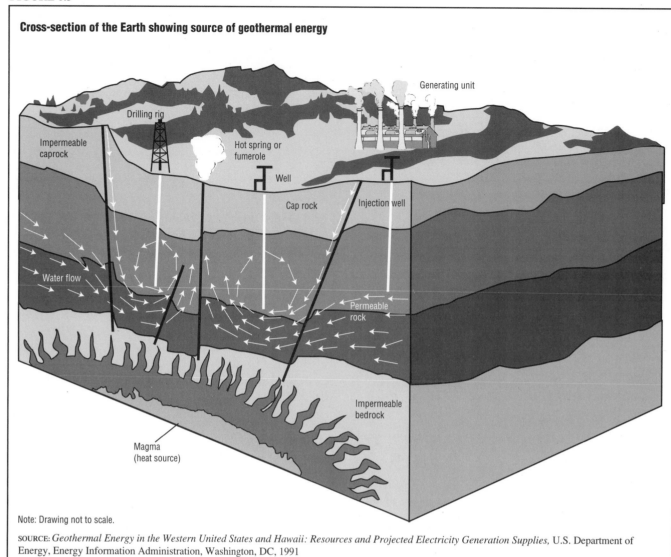

Cross-section of the Earth showing source of geothermal energy

Note: Drawing not to scale.

SOURCE: *Geothermal Energy in the Western United States and Hawaii: Resources and Projected Electricity Generation Supplies,* U.S. Department of Energy, Energy Information Administration, Washington, DC, 1991

GEOPRESSURIZED RESERVOIRS. Geopressurized reservoirs are sedimentary formations containing hot water and methane gas. Supplies of geopressurized energy remain uncertain, and drilling is expensive. Scientists hope that advancing technology will eventually permit the commercial exploitation of the methane content in these reservoirs.

MAGMA RESOURCES. Magma resources are found from 10,000 to 33,000 feet below the earth's surface, where molten or partially liquefied rock is located. Because magma is so hot, ranging from 1,650 to 2,200 degrees Fahrenheit, it is a good geothermal resource. The process for extracting energy from magma is still in the experimental stages.

Disadvantages of Geothermal Energy

Geothermal plants must be built near a geothermal source, are not very efficient, produce unpleasant odors from sulfur released in processing, generate noise, are inaccessible for most states, release potentially harmful pollutants (hydro-

gen sulfide, ammonia, and radon), and release poisonous arsenic or boron often found in geothermal waters. Serious environmental concerns have been raised over the release of chemical compounds, the potential contamination of water sources, the collapse of the land surface around the area from which the water is being drained, and potential water shortages resulting from massive withdrawals of water.

Domestic Production of Geothermal Energy

According to the EIA's *Annual Energy Review 2000* (2001), geothermal energy ranked third in renewable energy production in the United States in 2000, after biomass and hydroelectric energy. (See Figure 6.1.) Public-sector involvement in the geothermal industry began with the passage of the Geothermal Steam Act of 1970 (PL 91-581), which authorized the U.S. Department of the Interior to lease geothermal resources on federal lands.

According to the International Geothermal Association, in 2002 the United States had 28 percent of the

FIGURE 6.4

Known fields for geothermal resources

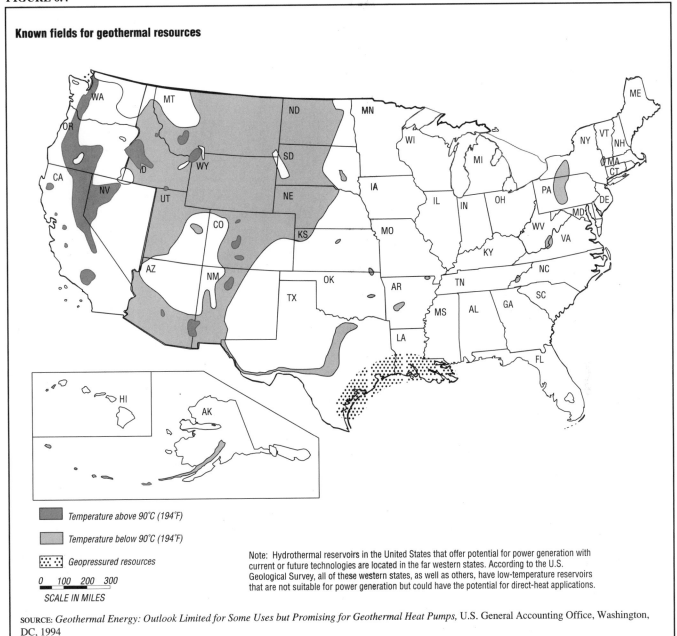

Temperature above 90°C (194°F)

Temperature below 90°C (194°F)

Geopressured resources

0   100   200   300
SCALE IN MILES

Note: Hydrothermal reservoirs in the United States that offer potential for power generation with current or future technologies are located in the far western states. According to the U.S. Geological Survey, all of these western states, as well as others, have low-temperature reservoirs that are not suitable for power generation but could have the potential for direct-heat applications.

SOURCE: *Geothermal Energy: Outlook Limited for Some Uses but Promising for Geothermal Heat Pumps,* U.S. General Accounting Office, Washington, DC, 1994

installed geothermal generating capacity of the world, but most of the easily exploited geothermal reserves in the United States have already been developed. In addition, utility companies and independent power producers are arguing over who should build additional generating capacity and what prices should be paid for the power. Continued growth in the American market depends on the regulatory environment, oil price trends, and the success of unproven technologies for economically exploiting some of the presently inaccessible geothermal reserves.

International Production of Geothermal Energy

Since 1970, worldwide geothermal electrical generating capacity has more than tripled. According to the *Annual Energy Review 2000,* geothermal energy made up about 1.5 percent of world electrical production in 1999. Nonetheless, geothermal sources worldwide produce little more than the energy output of ten average-size coal-fired power plants.

World geothermal reserves are immense but unevenly distributed. They fall mostly in seismically active areas at the margins or borders of the earth's nine tectonic plates. Currently, exploited reserves represent only a small fraction of the overall potential—many countries are believed to have in excess of 100,000 megawatts of geothermal energy available.

The Philippine government has committed itself to the development of geothermal power by providing tax incentives and cooperation to the private sector; as of 2000 geothermal energy provided about one-fourth of the nation's electricity. New Zealand and Iceland both use their rich steam reserves to provide significant amounts of power—in 2000 Iceland provided about 15 percent of its national energy total with geothermal energy. Indonesia, Italy, Japan, and Mexico also have significant geothermal capacity.

A few nations in the developing world—including El Salvador, Kenya, Bolivia, Costa Rica, Ethiopia, India, and Thailand—have considerable steam reserves available for power generation. Debt-ridden developing nations are especially eager to use available geothermal reserves instead of relying on fossil fuel imports for their energy needs.

WIND ENERGY

Winds are created by the uneven heating of the atmosphere by the Sun, the irregularities of the earth's surface, and the rotation of the earth. Winds are strongly influenced by local terrain, water bodies, weather patterns, vegetation, and other factors. Wind flow, when "harvested" by wind turbines, can be used to generate electricity.

Wind machines have changed dramatically from those that were common in the 1800s. Early windmills produced mechanical energy to pump water and run sawmills. In the late 1890s, Americans began experimenting with wind power to generate electricity. Their early efforts produced enough electricity to light one or two modern lightbulbs.

Compared to the pinwheel-shaped farm windmills that can still be seen dotting the American rural landscape, today's state-of-the-art wind turbines look more like airplane propellers. Their sleek, high-tech fiberglass design and aerodynamics allow them to generate an abundance of electricity while they also produce mechanical energy and heat.

Beginning in the 1990s, industrial and developing countries alike have started using wind power as a source of electricity to complement existing power sources and to bring electricity to remote regions. Wind turbines cost less to install per unit of kilowatt capacity than either coal or nuclear facilities. After installing a windmill, there are few additional costs, as the fuel (wind) is free.

Wind speeds are generally highest and most consistent in mountain passes and along coastlines. Europe has the greatest coastal wind resources, and clusters of wind turbines, or wind farms, are being developed there and in Asia. Denmark, the Netherlands, China, and India are especially interested in fostering the development of domestic wind industries. In the United States it is estimated that sufficient wind energy is available to provide more than 1 trillion kilowatt-hours of electricity annually, or about one-third of the total used in 1999. Currently, electricity-producing wind turbines operate in 95 countries.

Energy Production by Wind Turbines

Wind is the world's fastest-growing renewable energy source. Although wind power has not been adopted widely in the United States, U.S. companies export turbines to Spain, the Netherlands, Great Britain, India, and China.

In the United States, the wind industry began in California in 1981 with the erection of 144 relatively small turbines capable of generating a combined total of 7 megawatts of electricity. Within a year the number of turbines had increased 10 times, and by 1986 they had multiplied 100-fold. In *Annual Energy Outlook 2002*, the EIA projects wind power capacity in the United States to grow by nearly 300 percent from 2000 to 2020.

The 1980s saw an explosion of wind technology in California, where about 95 percent of the installed wind capacity in the United States used to be located. During 1998 and 1999, however, wind farm activity expanded into other states; less than 32 percent of new wind power construction was located in California. This increasing activity outside California was motivated by financial incentives (such as the wind energy production tax credit), regulatory incentives, and state mandates (in Iowa and Minnesota). In 1999 Iowa, Minnesota, and Texas each had capacity additions exceeding 100 megawatts. Other wind farm projects exist in Hawaii, Montana, New York, Oregon, and Wyoming. A large-scale project was approved in West Virginia in July 2002, but construction had not yet begun as of December 2002. Twelve states—North Dakota, South Dakota, Texas, Kansas, Montana, Nebraska, Wyoming, Oklahoma, Minnesota, Iowa, Colorado, and New Mexico—contain 90 percent of the U.S. wind energy potential.

Refinements in wind-turbine technology may enable a substantial portion of the nation's electricity to be produced by wind energy. Use of this technology is being encouraged by the initiative "Wind Powering America," which was announced in June 1999 by the U.S. Secretary of Energy. The stated goals of the program are to have 80,000 megawatts of wind power generation capacity in place by 2020 and to have wind power provide 5 percent of the nation's electricity generation.

An added incentive to developing wind technology is continuing tax credits. The wind-energy-production tax credit provided by the Energy Policy Act of 1992 (EPACT) was scheduled to expire in 1999 but was extended to the end of 2003.

International Development of Wind Energy

During the decade following the 1973 oil embargo, more than 10,000 wind machines were installed worldwide, ranging in size from portable units to multimegawatt turbines. In developing nations villages, small wind turbines recharge batteries and provide essential services. In China small wind turbines allow people to watch their favorite television shows, which has increased demand. In fact, in 2001 China was the world's largest manufacturer of small wind turbines.

Global wind-power-generating capacity was about 23,300 megawatts in 2001, up from 7,200 megawatts in 1997 and 3,000 megawatts in 1993. According to the EIA's *International Energy Outlook 2002* (2001), about 3,800 megawatts of new wind energy capacity were added worldwide during 2000. Germany, the country with the largest wind capacity, added 1,650 of these megawatts. Spain added 795 megawatts and Denmark 588 megawatts. Wind power remains the fastest growing renewable source of electricity. The largest producers of wind energy are Germany, the United States, Denmark, and India.

Interest in wind energy has been driven, in part, by the declining cost of capturing wind energy. From more than 25 cents per kilowatt-hour in 1980, wind energy prices declined to about 5 cents per kilowatt-hour in 2002 for new turbines at sites with strong winds. Decreasing costs could make wind power competitive with gas and coal power plants, even before considering wind's environmental advantages.

Advantages and Disadvantages of Wind Energy

The main problem with wind energy is that the wind does not always blow. Some people object to the whirring noise of wind turbines or do not like to see wind turbines clustered in mountain passes and along shorelines because they interfere with scenic views. Environmentalists have charged that wind turbines are responsible for the loss of thousands of endangered birds that fly into the blades, as birds frequently use windy passages in their travel patterns.

However, generating electricity with wind offers many environmental advantages. Wind farms do not emit climate-altering carbon dioxide, acid-rain-forming pollutants, respiratory irritants, or nuclear waste. Because wind farms do not require water to operate, they are especially well suited to semi-arid and arid regions.

SOLAR ENERGY

Ancient Greek and Chinese civilizations used glass and mirrors to direct the Sun's rays to start fires. Solar energy (energy from the Sun) is a renewable, widely available energy source that does not generate greenhouse gases or radioactive waste. Solar-powered cars have competed in long-distance races, and solar energy has been used routinely for many years to power spacecraft. Although many people consider solar energy a product of the space age, architectural researchers at the Massachusetts Institute of Technology built the first solar house in 1939.

Solar radiation is nearly constant outside the Earth's atmosphere, but the amount of solar energy reaching any point on Earth varies with changing atmospheric condi-

FIGURE 6.5

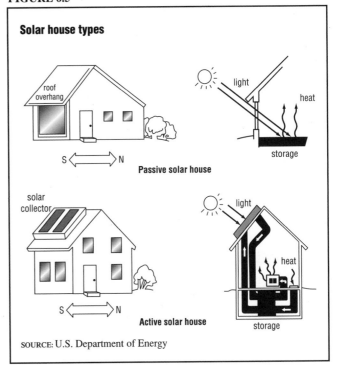

Solar house types

Passive solar house

Active solar house

SOURCE: U.S. Department of Energy

tions, such as clouds and dust, and the changing position of the Earth relative to the Sun. In the United States, exposure to the Sun's rays is greatest in the West and Southwest regions. Nevertheless, almost all U.S. regions have solar resources that can be used.

Passive and Active Solar Energy Collection Systems

Passive solar energy systems, such as greenhouses or windows with a southern exposure, use heat flow, evaporation, or other natural processes to collect and transfer heat. (See Figure 6.5.) They are considered the least costly and least difficult solar systems to implement.

Active solar systems use mechanical methods to control the energy process. (See Figure 6.5.) They require collectors and storage devices as well as motors, pumps, and valves to operate the systems that transfer heat. Collectors consist of an absorbing plate that transfers the Sun's heat to a working fluid (liquid or gas), a translucent cover plate that prevents the heat from radiating back into the atmosphere, and insulation on the back of the collector panel to further reduce heat loss. Excess solar energy is transferred to a storage facility so it may be used to provide power on cloudy days. In both active and passive systems, the conversion of solar energy into a form of power is made at the site where it is used. The most common and least expensive active solar systems are used for heating water.

Solar Thermal Energy Systems

A solar thermal energy system uses intensified sunlight to heat water or other fluids to temperatures of more than 750 degrees Fahrenheit. Mirrors or lenses constantly track the Sun's position and focus its rays onto solar receivers that contain fluid. Solar heat is transferred to the water, which in turn powers an electric generator. In a distributed solar thermal system, the collected energy powers irrigation pumps, providing electricity for small communities or capturing normally wasted heat from the Sun in industrial areas. In a central solar thermal system, the energy is collected at a central location and used by utility networks for a large number of customers.

Other solar thermal energy systems include solar ponds and trough systems. Solar ponds are lined ponds filled with water and salt. Because salt water is denser than fresh water, the salt water on the bottom absorbs the heat, and the fresh water on top keeps the salt water contained and traps the heat. Trough systems use U-shaped mirrors to concentrate the sunshine on water or oil-filled tubes.

Photovoltaic Conversion Systems

The photovoltaic (PV) cell solar energy system converts sunlight directly into electricity without the use of mechanical generators. PV cells have no moving parts, are easy to install, require little maintenance, do not pollute the air, and can last up to 20 years. PV cells are commonly used to power small devices, such as watches or calculators. They are also being used on a larger scale to provide electricity for rural households, recreational vehicles, and businesses. Solar panels using photovoltaic cells have generated electricity for space stations and satellites for many years.

Since PV systems produce electricity only when the sun is shining, a backup energy supply is required. PV cells produce the most power around noon, when sunlight is the most intense. A photovoltaic system typically includes storage batteries that provide electricity during cloudy days and at night.

The use of photovoltaic technology is expanding both in the United States and abroad. PV systems have low operating costs because there are no turbines or other moving parts, and maintenance is minimal. PV cell systems are nonpolluting and silent and can be operated by computer. Above all, the fuel source (sunshine) is free and plentiful. The main disadvantage of photovoltaic cell energy systems is the initial cost. Although the price has fallen considerably, PV cells are still too expensive for widespread use. PV systems also use some toxic materials, which may cause environmental problems.

Solar Energy Usage

Because it is difficult to measure directly the use of solar energy, shipments of solar equipment can be used as an indicator of use. From a high of 84 low-temperature solar collector manufacturers in 1979, this number dropped to only 13 manufacturers in 1999. Total shipments of solar thermal collectors peaked in 1981 at more than 21 million square feet and declined to approximately 8.6 million square feet by 1999. (See Figure 6.6.)

Most of the solar thermal collectors sold are for residential purposes (mostly in the Sunbelt states). (See Figure 6.7.) In 1999 most of the solar thermal collectors shipped were used for heating swimming pools, and a smaller percentage for hot water. (See Figure 6.8.) The market for solar energy space heating has virtually disappeared. Only a small proportion of solar thermal collectors are used for commercial purposes, though some state and municipal power companies have added solar energy systems as adjuncts to their regular power sources during peak hours.

Solar Power as an International Rural Solution

Rural areas are more expensive to serve with energy than cities, and electrification has been slow to reach many people in rural areas of developing countries. In the United States, it was only in 1935, after the Rural Electrification Administration provided low-cost financing to rural electric cooperatives, that most farmers received power. In places such as western China, the Himalayan foothills, and the Amazon basin, the cost of connecting new rural customers to electricity grids is much higher than that in cities. Furthermore, state-owned power systems have been poorly managed in many countries. This has left many national power systems all but bankrupt, and blackouts have become common.

In India blackouts are so common that many factories and other businesses have, at great expense, set up their own private systems, using natural gas, propane, or fuel oil. Although rural families do not have access to those systems, they do have sunlight. In most tropical countries, considerable sunlight falls on rooftops. Electricity produced by solar photovoltaic cells was initially too expensive, as much as a thousand times more than that from conventional plants, but it has continually fallen in price.

During the 1990s a different approach to solar electrification developed, driven less by government planners and more by the desire of individual families to meet their own needs for electricity. In more than a dozen countries, solar power is reaching thousands of families one by one, avoiding the delays of government planners. In Kenya, where the state power company was on the verge of bankruptcy, eight domestic companies merged to market, install, and maintain solar home systems. With little state or international assistance, these companies managed to electrify 20,000 rural households between 1987 and 1992, 3,000 more than the state power system.

Internationally, solar photovoltaics experienced the largest growth rate of any renewable energy product in 2000, according to *Renewables Information 2002* (International Energy Agency). Germany and Korea were the largest producers of photovoltaics, but production

FIGURE 6.6

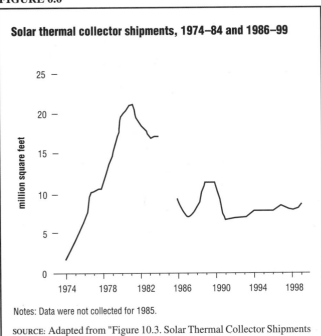

Solar thermal collector shipments, 1974–84 and 1986–99

Notes: Data were not collected for 1985.

SOURCE: Adapted from "Figure 10.3. Solar Thermal Collector Shipments by Type, Price, and Trade," in *Annual Energy Review 2000*, U.S. Department of Energy, Energy Information Administration, Washington, DC, 2001.

FIGURE 6.7

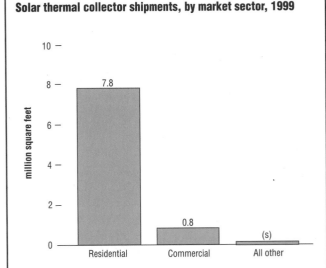

Solar thermal collector shipments, by market sector, 1999

(s)=Less than 0.05 million square feet.

SOURCE: Adapted from "Figure 10.4. Solar Thermal Collector Shipments by End Use, Market Sector, and Type, 1999," in *Annual Energy Review 2000*, U.S. Department of Energy, Energy Information Administration, Washington, DC, 2001.

FIGURE 6.8

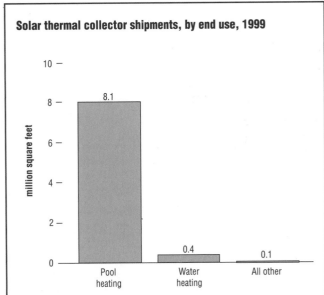

Solar thermal collector shipments, by end use, 1999

SOURCE: Adapted from "Figure 10.4. Solar Thermal Collector Shipments by End Use, Market Sector, and Type, 1999," in *Annual Energy Review 2000*, U.S. Department of Energy, Energy Information Administration, Washington, DC, 2001.

occurred in many countries. Solar thermal energy use made small gains. This type of energy supply is concentrated mainly in the United States, Japan, and Turkey.

Advantages and Disadvantages of Solar Energy

The primary advantage of solar energy is its inexhaustible supply, while its primary disadvantage is its reliance on a consistently sunny climate to provide continuous electrical power, which is only possible in limited areas. In addition, a large amount of land area is necessary for the most efficient collection of solar energy by electricity plants. Experts estimate that a new thermal energy plant would have a 60 percent higher cost of production than a conventional coal-fired plant.

Future Development Trends

Interest in photovoltaic solar energy systems is particularly high in rural and remote areas where it is impractical to extend traditional electrical power lines. In some remote areas, PV cells are used as independent power sources for communications or for the operation of water pumps or refrigerators.

Although solar power still costs more than three times as much as fossil fuel energy, utilities could turn to solar energy to provide "peaking power" on extremely hot or cold days. In the long run, some people believe that building solar energy systems to provide peak power capacity would be cheaper than building the new and expensive diesel fuel generators that are now used.

POWER FROM THE OCEAN

The potential power of the world's oceans is unknown. Because the ocean is not as easily controlled as a river or water that is directed through canals into turbines, unlocking that potential power is far more challenging. Three ideas being considered are tidal plants, wave power, and ocean thermal energy conversion (OTEC).

Tidal Power

The tidal plant uses the power generated by the tidal flow of water as it ebbs, or flows back out to sea. A minimum tidal range of three to five yards is generally considered necessary for an economically feasible plant. (The tidal range is the difference in height between consecutive high and low tides.) Canada has built a small 40-megawatt unit at the Bay of Fundy, with its 15-yard tidal range, the largest range in the world, and is considering building a larger unit there. The largest existing tidal facility is the 240-megawatt plant at the La Rance estuary in northern France, built in 1965. Russia has a small 400-kilowatt plant near Murmansk, close to the Barents Sea.

Wave Energy

Norway has two operating wave power stations at Tostestallen on its Atlantic coast. The arrival of a wave forces water up a hollow 65-foot tower, displacing the air already in the tower. This air rushes out of the top through a turbine. The rotors of that turbine then spin, generating electricity. When the wave falls back and the water level falls, air is sucked back in through the turbine, again generating electricity.

A second type of wave energy power plant uses the overflow of high waves. As the wave splashes against the top of a dam, some of the water goes over and is trapped in a reservoir on the other side. The water is then directed through a turbine as it flows back to the sea.

These two kinds of plants are experimental. Several projects are under way in Japan and the Pacific region to determine a way to use the potential of the huge waves of the Pacific. Although considerable progress has been made in the research and development of this technology, several challenging engineering problems remain to be solved.

Ocean Thermal Energy Conversion (OTEC)

Ocean thermal energy conversion, or OTEC, uses the temperature difference between the ocean's warm surface water and the cooler water in its depths to produce heat energy that can power a heat engine to produce electricity. OTEC systems can be installed on ships, barges, or offshore platforms with underwater cables that transmit electricity to shore.

HYDROGEN: A FUEL OF THE FUTURE?

Hydrogen, the lightest and most abundant chemical element, is the ideal fuel from the environmental point of view. Its combustion produces only water vapor, and it is entirely carbon-free. Three-quarters of the mass of the universe is hydrogen. However, the combustible form of hydrogen is a gas and is not found in nature. The many compounds containing hydrogen, for example water, cannot be converted into pure hydrogen without the expenditure of energy. The amount of energy that would be required to make gas is about the same as the amount of energy that would be obtained by the combustion of the hydrogen. Therefore, with today's technology, little or nothing could be gained from an energy point of view.

Scientists, however, are researching ways to produce hydrogen economically. Whether this will come from fusion, solar energy, or elsewhere is not possible to predict now. Scientists have considered the possibility of a transition to hydrogen for more than a century, and today many see hydrogen as the logical "third-wave" fuel, with hydrogen gas following oil, just as oil replaced coal decades earlier. For now, however, widespread use of hydrogen as fuel is purely theoretical.

FINANCIAL OBSTACLES TO INTERNATIONAL RENEWABLE ENERGY USE

Despite technological advances, huge numbers of people worldwide have no access to electricity. Part of the problem is lack of credit. Consumer credit is one of the most momentous financial advances of the twentieth century, leading to wide ownership of homes, automobiles, and appliances that the average person could not afford outright. Millions of families in the developing world might afford solar energy if credit were available.

To fill this gap, international agencies are working to set up revolving credit funds to finance solar, wind, and other renewable energy systems in many countries, including Vietnam, India, Indonesia, Uganda, Swaziland, and the Dominican Republic. A number of nonprofit agencies, private foundations, investment firms, and the World Bank have become involved. Many observers hope that newly developing nations can avoid many of the damaging environmental practices of industrialization, including a heavy dependence on fossil fuels.

FUTURE TRENDS IN U.S. RENEWABLE ENERGY USE

In *Annual Energy Outlook 2002* (2001), the EIA forecasts that for renewable energy sources, including ethanol for transportation, consumption will increase by 1.7 percent per year from 2000 to 2020. Use of renewable fuels for electricity generation is also projected to grow by 1.7 percent per year. In 2000 renewable energy provided 357 billion kilowatt-hours of electricity, which is projected to increase to 464 billion kilowatt-hours in 2020. Nevertheless, since electricity generation in general will grow from 2000 to 2020, renewable energy is expected to keep a steady 9 percent share of total electricity generation during that time.

Hydropower production is projected to rise only slightly from 276 billion kilowatt-hours in 2000 to 304 billion kilowatt-hours in 2020. The production of other renewables should increase steadily. (See Figure 6.9.) For example, significant increases are projected for both geothermal energy

FIGURE 6.9

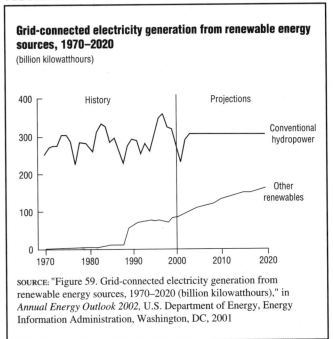

Grid-connected electricity generation from renewable energy sources, 1970–2020

(billion kilowatthours)

SOURCE: "Figure 59. Grid-connected electricity generation from renewable energy sources, 1970–2020 (billion kilowatthours)," in *Annual Energy Outlook 2002,* U.S. Department of Energy, Energy Information Administration, Washington, DC, 2001

FIGURE 6.10

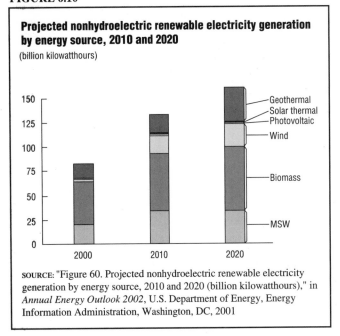

Projected nonhydroelectric renewable electricity generation by energy source, 2010 and 2020

(billion kilowatthours)

SOURCE: "Figure 60. Projected nonhydroelectric renewable electricity generation by energy source, 2010 and 2020 (billion kilowatthours)," in *Annual Energy Outlook 2002,* U.S. Department of Energy, Energy Information Administration, Washington, DC, 2001

and wind power capacity from 2000 to 2020. The EIA projects that high-output geothermal capacity will increase by 87 percent to 5 gigawatts and is forecasted to provide 1 percent of the total energy supply in the United States by 2020. Wind capacity will likely increase by 300 percent, from 5 billion kilowatt-hours in 2000 to 24 billion kilowatt-hours in 2020. This, however, will provide less than 1 percent of the nation's electricity needs.

Municipal solid waste energy production is expected to increase by 11 billion kilowatt-hours from 2000 to 2020. The largest source of renewable generation (not including hydropower) is biomass, which is projected to increase from 38 billion kilowatt-hours in 2000 to 64 billion kilowatt-hours in 2020—a 74 percent increase. This comprises 1 percent of the total electricity supply of the United States. (See Figure 6.10.) Solar energy is not expected to contribute much to centrally generated electricity.

CHAPTER 7
ENERGY RESERVES—OIL, GAS, COAL, AND URANIUM

Fossil fuels are nonrenewable resources. Nonrenewable resources are defined as concentrations of solid, liquid, or gaseous hydrocarbons that occur naturally in or near the earth's surface. These resources must be currently or potentially recoverable for economic use. Once used, these substances cannot be replaced. Knowing estimates of the recoverable quantities of crude oil, natural gas, coal, and uranium resources in the United States and worldwide is essential to the development, implementation, and evaluation of national energy policies and legislation. In the United States, Congress requires the Department of Energy (DOE) to prepare estimates of energy reserves.

Proved reserves are reserves from known locations that geological and engineering data demonstrate, with reasonable certainty, to be recoverable with current technological means and economic conditions. Undiscovered recoverable resources are quantities of fuel that are thought to exist in favorable geologic settings. These resources would be feasible to retrieve with existing technological means, although they may not be feasible to recover under current economic conditions.

CRUDE OIL

During the decade from 1990 through 2000, total U.S. crude oil proved reserves declined through 1996, then fluctuated from 1997 through 2000. (See Figure 7.1.) On December 31, 2000, crude oil reserves were at 22 billion barrels—about the 1996 reserves level. Together, Texas, Alaska, California, and the Gulf of Mexico offshore areas account for 77 percent of U.S. proved reserves. (See Table 7.1.) Of these four regions, only the Gulf of Mexico had an increase in crude oil proved reserves in 2000.

Proved reserves of crude oil and natural gas rose in 1970 with the inclusion of Alaska's North Slope oil fields. Since then, Alaskan reserves have steadily declined. In 1987 Alaska was estimated to have 13.2 billion barrels of crude oil; by 2000 it had only 4.9 billion barrels. (See Fig-

TABLE 7.1

Proved reserves of crude oil as of December 31, 2000, by selected states

Area	Percent of U.S. Oil Reserves
Texas	24
Alaska	22
California	17
Gulf of Mexico Federal Offshore	14
Area Total	**77**

SOURCE: *U.S. Crude Oil, Natural Gas, and Natural Gas Liquids Reserves: 2000 Annual Report,* U.S. Department of Energy, Energy Information Administration, Washington, DC, 2001

ure 7.2.) Crude oil proved reserves fell by 39 million barrels in Alaska from 1999 to 2000. The discovery of a new field in 1996, the Arco-owned Alpine field on Alaska's North Slope, was expected to raise recovery by 70,000 barrels per day by 2001. The Alpine field went into production in November 2001 and is performing better than expected.

The Gulf of Mexico Federal Offshore areas, which are in U.S. territorial waters, had about 3.2 billion barrels of crude oil proved reserves in 2000, up 430 million barrels from 1999. (See Figure 7.2). Improvements in deepwater drilling systems—floating platforms and subsea wells—have allowed the industry to expand into continually deeper Gulf waters in search of crude oil. The Gulf holds much promise for future reserves discoveries.

In 1996 scientists turned their attention to the Permian Basin of west Texas and southeastern New Mexico, where plenty of dry-land potential for crude oil was found, making it one of the most active onshore areas for recent exploration. Still, in 2000 Texas experienced the largest decline in reserves of any state except Louisiana. The Texas decline, though large, represented only 2 percent of Texas's 1999 reserves.

FIGURE 7.1

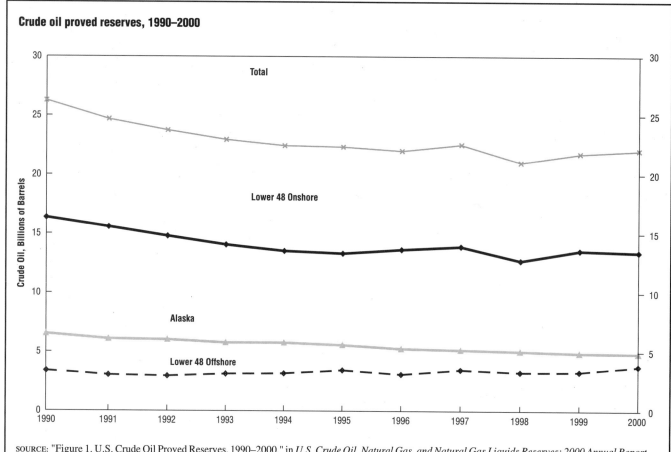

Crude oil proved reserves, 1990–2000

SOURCE: "Figure 1. U.S. Crude Oil Proved Reserves, 1990–2000," in *U.S. Crude Oil, Natural Gas, and Natural Gas Liquids Reserves: 2000 Annual Report,* U.S. Department of Energy, Energy Information Administration, Washington, DC, 2001

NATURAL GAS

Figure 7.3 shows that in 2000 the United States had about 177 trillion cubic feet of dry natural gas proved reserves. Additionally the United States had 8.3 billion barrels of natural gas liquid proved reserves. This is an increase in natural gas reserves from 1999.

UNDISCOVERED OIL AND GAS RESOURCES

In addition to those proved resources, other resources are believed to exist based on past geological experience, although they are not yet proved. For 2000 the DOE's Energy Information Administration (EIA) reported there were an estimated 105 billion barrels of crude oil, 682 trillion cubic feet of natural gas, and 8 billion barrels of dry natural gas liquids as undiscovered resources in the United States. (See Table 7.2, where these resources are called "technically recoverable resources in U.S. undiscovered conventionally reservoired fields.")

Looking for Oil and Gas

Finding oil and gas is usually a two-step process. First, geological and geophysical exploration identifies the areas where oil and gas may most likely be found. Much of this exploration is seismic, which measures the movement of the earth. Then exploratory wells are drilled to determine if oil or gas is present.

Market conditions and technological developments shape exploration for oil and gas. For example, the economic problems of the oil industry can be seen in the drop in the number of exploratory oil and gas wells completed. Drilling activity for exploratory wells has declined dramatically since the early 1980s. (See Table 7.3.) In 1981 a peak of 91,550 exploratory wells were drilled, with nearly 70 percent of those successful. In 2000 only 25,140 were attempted, but 79.3 percent were successful. In 1981, 3,970 rotary rigs were in operation; by 2000 only 918 were operating. (See Figure 7.4.) Of.this number, 721 rigs drilled for gas, while 197 drilled for oil. There were 778 onshore rigs and 140 offshore rigs. The average depth of exploratory and development wells has steadily increased, from 3,635 feet in 1949 to 6,516 feet in 2000. Gas wells (averaging 6,865 feet in 2000) are typically deeper than oil wells (5,534 feet).

The Cost to Drill

In 1999 an average well cost about $856,100 to drill, or $152.02 per foot. (See Table 7.4.) A gas well generally

FIGURE 7.2

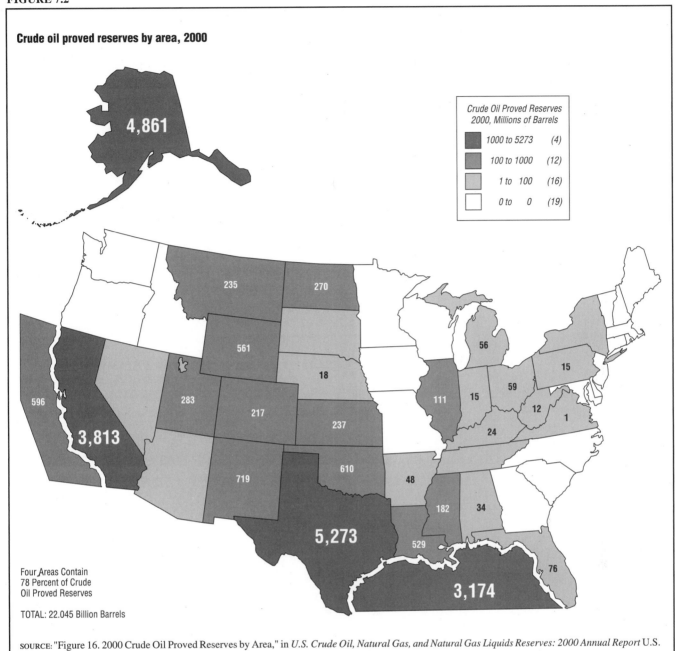

Crude oil proved reserves by area, 2000

Crude Oil Proved Reserves
2000, Millions of Barrels

■	1000 to 5273	(4)
■	100 to 1000	(12)
■	1 to 100	(16)
□	0 to 0	(19)

Four Areas Contain
78 Percent of Crude
Oil Proved Reserves

TOTAL: 22.045 Billion Barrels

SOURCE: "Figure 16. 2000 Crude Oil Proved Reserves by Area," in *U.S. Crude Oil, Natural Gas, and Natural Gas Liquids Reserves: 2000 Annual Report* U.S. Department of Energy, Energy Information Administration, Washington, DC, 2001

costs more than an oil well to drill because it is deeper. In 1999, however, the cost of drilling an average gas well ($798,400) was not much higher than that of drilling an oil well ($783,000), because the average cost per foot of drilling an oil well ($156.45) was, for the first time, higher than that of drilling a foot of gas well ($138.42). Although drilling costs have fluctuated in recent years, it costs considerably more to drill a well today than it did in the 1960s and early 1970s, not only because of inflation but also because wells must now be drilled deeper. Even in as short a period as 1996 to 1999, the cost per well increased significantly, at nearly $360,000 more per well, or an increase of more than 75 percent.

Spending on Exploration and Development

The estimated expenditures on exploration for, and development of, oil and gas by major U.S. energy-producing companies peaked at $65.3 billion in 1984. (See Table 7.5.) U.S. energy companies spent $31.3 billion in 1999 on oil and gas exploration around the world, with about 43 percent of that, $13.5 billion, spent for exploration in the United States.

The chance of finding a major oil or gas field has become increasingly smaller. Oil explorers are spending more of their time looking for less oil. This does not necessarily imply that the nation will soon be facing a shortage of

FIGURE 7.3

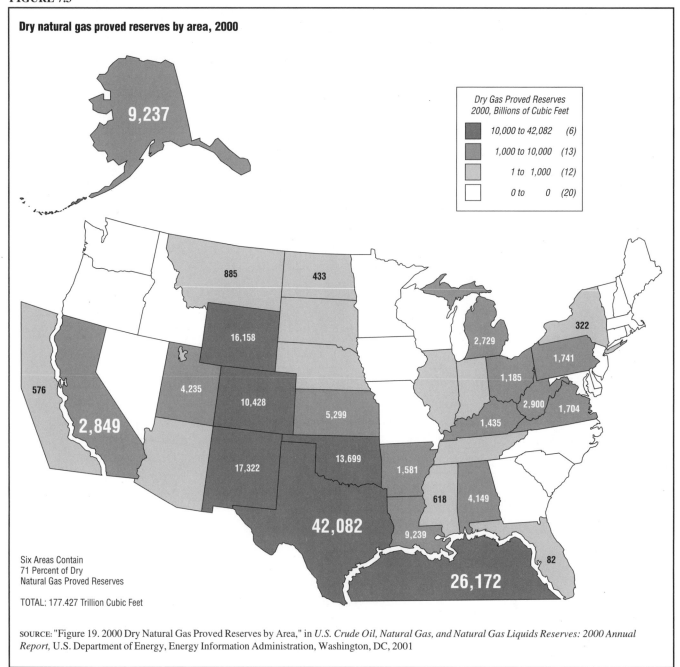

Dry natural gas proved reserves by area, 2000

Dry Gas Proved Reserves
2000, Billions of Cubic Feet

▨	10,000 to 42,082	(6)
▨	1,000 to 10,000	(13)
▨	1 to 1,000	(12)
☐	0 to 0	(20)

Six Areas Contain
71 Percent of Dry
Natural Gas Proved Reserves

TOTAL: 177.427 Trillion Cubic Feet

SOURCE: "Figure 19. 2000 Dry Natural Gas Proved Reserves by Area," in *U.S. Crude Oil, Natural Gas, and Natural Gas Liquids Reserves: 2000 Annual Report,* U.S. Department of Energy, Energy Information Administration, Washington, DC, 2001

gas or oil; reserves are still considerable, especially of natural gas. It does show, however, that the oil and gas industries are mature, or highly developed. Finding new resources will become more expensive, and until the prices of oil and gas rise sharply, fewer companies will go looking for it.

Drilling in the Arctic National Wildlife Refuge (ANWR)

Much controversy developed over opening the Arctic National Wildlife Refuge (ANWR) in Alaska to oil drilling. Prudhoe Bay, directly to the west of the refuge, supplies about 60 percent of Alaska's oil and 20 percent of the country's domestic oil, although production is dropping steadily as the oil is used up. In 1980 Congress passed the Alaska National Interest Lands Conservation

Act (PL 96-487), which set aside more than 104 million acres for parks, refuges, and wilderness areas, including ANWR (18 million acres), but the conservation act did not include the coastal plain.

The U.S. Department of the Interior (DOI), in a 1987 report to Congress, recommended that the 1.5-million-acre coastal plain be opened for exploration and extraction. Alaskan corporations supported the proposal in hopes of sharing in the proceeds. Environmentalists strongly opposed the plan because of the destruction that drilling could do to native wildlife, such as caribou, polar and grizzly bears, musk oxen, wolves, Arctic foxes, and millions of nesting birds.

TABLE 7.2

Mean estimates of technically recoverable oil and gas resources by deposit type and location, 2000

Area	Jurisdiction	Crude oil (billion barrels)	Natural gas (dry) (trillion cubic feet)	Natural gas liquids (billion barrels)
Undiscovered Conventionally Reservoired Fields				
Alaska onshore + state offshore	Federal	3.75	33.97	0.54
Alaska onshore + state offshore	Other	4.68	95.37	0.61
Alaska federal offshore	Federal	24.90	122.60	0.00
Lower 48 states onshore + state offshore	Federal	3.79	23.97	1.26
Lower 48 states onshore + state offshore	Other	17.83	166.41	5.64
Lower 48 states federal offshore	Federal	50.10	239.60	0.00
Alaska subtotal		33.33	251.94	1.15
Alaska percentage federal		86.0%	62.1%	47.0%
Lower 48 states subtotal		71.72	429.98	6.90
Lower 48 states percentage federal		75.1%	61.3%	18.3%
Technically Recoverable Resources in U.S. Undiscovered Conventionally Reservoired Fields		**105.05**	**681.92**	**8.05**
Percentage Federal		**78.6%**	**61.6%**	**22.4%**
Ultimate Recovery Appreciation				
U.S. onshore + state offshore	Federal	14.33	118.70	4.94
U.S. onshore + state offshore	Other	45.67	203.30	8.46
U.S. federal offshore	Federal	7.70	68.00	0.00
Technically Recoverable Resources in U.S. from Ultimate Recovery Appreciation in Discovered Conventionally Reservoired Fields		**67.70**	**390.00**	**13.40**
U.S. Percentage Federal		**32.5%**	**47.9%**	**36.9%**
Continuous Type Deposits				
Non-coal bed	Federal	0.32	127.08	1.45
Non-coal bed	Other	1.75	181.72	0.67
Coal bed	Federal	0.00	16.08	0.00
Coal bed	Other	0.00	33.83	0.00
Non-coal bed subtotal		2.07	308.80	2.12
Non-coal bed percentage federal		15.5%	41.2%	68.4%
Coal bed subtotal		0.00	49.91	0.00
Coal bed percentage federal		0.0%	32.2%	0.0%
Technically Recoverable Resources in U.S. from Continuous Type Deposits		**2.07**	**358.71**	**2.12**
Continuous Type Percentage Federal		**15.5%**	**39.9%**	**68.4%**
U.S. Totals All Sources				
U.S. onshore + state offshore	Federal	22.19	319.80	8.19
U.S. onshore + state offshore	Other	69.93	680.63	15.38
Federal offshore	Federal	82.70	430.20	0.00
Federal subtotal		104.89	750.00	8.19
U.S. Technically Recoverable Resources		**174.82**	**1,430.63**	**23.57**
Percentage Federal		**60.0%**	**52.4%**	**34.7%**

Notes: Proved Reserves are not included in these estimates.
Federal Onshore excludes Indian and native lands even when federally managed in trust.
Zero (0) indicates either that none exists in this area or that no estimate of this resource has been made for this area.
The estimates of ultimate recovery appreciation for onshore and state offshore lands were imputed by assuming that the total estimates thereof reported by the U.S. Geological Survey (USGS) could be apportioned according to the ratio of 1996 production from onshore federal lands to total U.S. production.
Federal offshore indicates MMS (Minerals Management Services) estimates for federal offshore jurisdictions (Outer Continental Shelf and deeper water areas seaward of state offshore). Probable and Possible reserves are considered by USGS definition to be part of USGS Reserve Growth but are separately considered by the MMS as its Unproved Reserves term. The USGS did not set a time limit for the duration of Reserve Growth; the MMS set the year 2020 as the time limit in its estimates of Reserve Growth in existing fields of the Gulf of Mexico. Excluded from the estimates are undiscovered oil resources in tar deposits and oil shales, and undiscovered gas resources in geopressured brines and gas hydrates.

SOURCE: "Table G1. Mean Estimates of Technically Recoverable Oil and Gas Resources by Deposit Type and Location," in *U.S. Crude Oil, Natural Gas, and Natural Gas Liquids Reserves: 2000 Annual Report*, U.S. Department of Energy, Energy Information Administration, Washington, DC, 2001

The DOI's Fish and Wildlife Service estimated that 3.2 billion barrels of recoverable oil exist in the contested area. They also estimated that there was a 46 percent chance of recovering the oil, a high figure by industry standards. Considering the general decline in American production, this could account for a significant portion of American petroleum production in this century. Vast quantities of natural gas are also likely to be found in the area.

The report further stated, however, that there could be a major effect on the migratory caribou herds, which number about 180,000 animals. While environmentalists estimated that 20 to 40 percent of the animals would be threatened, DOI officials claimed that the caribou would change their migratory habits.

If major oil reserves are found, oil companies might operate on the coastal plain for several decades. Possible

TABLE 7.3

Oil and gas exploratory and developmental wells, 1949–2000

Year	Wells drilled (thousands)				Successful wells (percent)	Footage drilled (million feet)				Average depth (feet per well)			
	Oil	Gas	Dry holes	Total		Oil	Gas	Dry holes	Total	Oil	Gas	Dry holes	Total
1949	21.35	3.36	12.60	37.31	66.2	79.4	12.4	43.8	135.6	3,720	3,698	3,473	3,635
1950	23.81	3.44	14.80	42.05	64.8	92.7	13.7	51.0	157.4	3,893	3,979	3,445	3,742
1951	23.18	3.44	17.03	43.64	61.0	95.1	13.9	63.1	172.1	4,103	4,056	3,706	3,944
1952	23.29	3.51	17.76	44.56	60.1	98.1	15.3	70.7	184.1	4,214	4,342	3,983	4,132
1953	25.32	3.97	18.45	47.74	61.4	102.1	18.2	73.9	194.2	4,033	4,599	4,004	4,069
1954	28.14	4.04	18.93	51.11	63.0	113.4	18.9	75.8	208.0	4,028	4,670	4,004	4,070
1955	30.43	4.27	20.45	55.15	62.9	121.1	19.9	85.1	226.2	3,981	4,672	4,161	4,101
1956	30.53	4.53	22.11	57.17	61.3	120.4	22.7	90.2	233.3	3,942	5,018	4,079	4,080
1957	27.36	4.48	20.16	52.00	61.2	110.0	23.8	83.2	217.0	4,021	5,326	4,126	4,174
1958	23.77	5.01	18.16	46.94	61.3	93.1	25.6	74.6	193.3	3,916	5,106	4,110	4,118
1959	24.04	4.93	18.59	47.56	60.9	94.6	26.6	79.5	200.7	3,935	5,396	4,275	4,220
1960	22.26	5.15	18.21	45.62	60.1	86.6	28.2	77.4	192.2	3,889	5,486	4,248	4,213
1961	21.44	5.49	17.33	44.25	60.8	85.6	29.3	74.7	189.6	3,994	5,339	4,311	4,285
1962	21.73	5.35	17.08	44.16	61.3	88.4	28.9	77.3	194.6	4,070	5,408	4,524	4,408
1963	20.14	4.57	16.76	41.47	59.6	81.8	24.5	76.3	182.6	4,063	5,368	4,552	4,405
1964	19.91	4.69	17.69	42.29	58.2	80.5	25.6	81.4	187.4	4,042	5,453	4,598	4,431
1965	18.07	4.48	16.23	38.77	58.2	73.3	24.9	76.6	174.9	4,059	5,562	4,723	4,510
1966	16.78	4.38	15.23	36.38	58.1	67.3	25.9	69.6	162.9	4,013	5,928	4,573	4,478
1967	15.33	3.66	13.25	32.23	58.9	58.6	21.6	61.1	141.4	3,825	5,898	4,616	4,385
1968	14.33	3.46	12.81	30.60	58.1	59.5	20.7	64.7	145.0	4,153	5,994	5,053	4,738
1969	14.37	4.08	13.74	32.19	57.3	61.6	24.2	71.4	157.1	4,286	5,918	5,195	4,881
1970	12.97	4.01	11.03	28.01	60.6	56.9	23.6	58.1	138.6	4,385	5,880	5,265	4,943
1971	11.85	3.97	10.31	26.13	60.6	49.1	23.5	54.7	127.3	4,126	5,890	5,305	4,858
1972	11.38	5.44	10.89	27.71	60.7	49.3	30.0	58.6	137.8	4,330	5,516	5,377	4,974
1973	10.17	6.93	10.32	27.42	62.4	44.4	38.0	55.8	138.2	R4,369	5,488	R5,403	5,041
1974	13.65	7.14	12.12	32.90	63.2	52.0	38.4	62.9	153.4	R3,812	5,387	R5,191	4,662
1975	16.95	8.13	13.65	38.72	64.8	66.8	R44.5	R69.2	180.5	R3,943	5,470	R5,073	4,661
1976	17.69	9.41	13.76	40.86	66.3	R68.9	49.1	69.0	187.0	R3,895	5,220	R5,014	4,577
1977	18.75	12.12	14.99	45.85	67.3	75.5	R63.7	R76.7	215.9	4,025	5,254	R5,120	4,708
1978	19.18	14.41	16.55	50.15	67.0	77.0	75.8	85.8	238.7	4,016	5,262	R5,183	4,760
1979	20.85	15.25	16.10	52.20	69.2	R82.7	80.5	R81.6	244.8	R3,966	5,275	R5,071	4,689
1980	32.64	17.33	20.64	70.61	70.8	124.3	91.5	98.9	314.7	R3,809	5,278	R4,790	4,456
1981	43.60	20.17	27.79	91.55	69.6	R171.2	107.8	R134.1	413.1	R3,926	5,346	R4,827	4,512
1982	39.20	18.98	26.22	84.40	68.9	148.8	106.7	122.8	378.3	R3,796	5,622	R4,683	4,482
1983	37.12	14.56	24.15	75.84	68.2	136.1	77.6	104.3	318.0	3,667	5,325	4,320	4,193
1984	42.61	17.13	25.68	85.41	69.9	161.8	90.6	119.0	371.4	R3,798	5,289	R4,634	4,348
1985	35.12	14.17	21.06	70.34	70.1	R137.4	R75.9	R99.8	313.0	3,911	5,354	R4,741	4,450
1986	19.10	8.52	12.68	40.29	68.5	76.6	44.7	60.5	181.9	4,013	5,253	R4,771	4,514
1987	16.16	8.06	11.11	35.33	68.5	66.3	42.5	53.4	162.2	R4,103	5,275	4,803	4,590
1988	13.64	8.56	10.04	32.23	68.8	58.7	45.3	52.3	156.4	4,305	5,298	5,211	4,851

TABLE 7.3

Oil and gas exploratory and developmental wells, 1949–2000 [CONTINUED]

Year	Wells drilled (thousands)				Successful wells (percent)	Footage drilled (million feet)				Average depth (feet per well)			
	Oil	Gas	Dry holes	Total		Oil	Gas	Dry holes	Total	Oil	Gas	Dry holes	Total
1989	10.20	9.54	8.19	27.93	70.7	43.3	49.2	R42.0	134.4	R4,244	R5,156	R5,124	4,813
1990	12.20	11.04	8.31	31.56	73.7	R54.5	56.1	43.1	153.7	R4,465	R5,083	R5,184	4,871
1991	11.77	9.53	7.60	28.89	73.7	R54.2	50.0	38.9	143.0	R4,601	R5,248	R5,118	4,950
1992	8.76	8.21	6.12	23.08	73.5	R43.9	46.0	31.2	121.1	R5,015	R5,608	R5,094	5,247
1993	8.41	10.02	6.33	24.75	74.4	42.5	60.0	32.6	135.1	R5,055	R5,993	5,150	5,459
1994	6.72	9.54	5.31	21.57	75.4	36.0	59.6	29.2	124.8	R5,356	R6,247	R5,506	5,787
1995	7.63	8.35	5.08	21.06	75.9	38.2	R51.3	R28.3	117.8	R5,009	R6,139	R5,585	5,596
1996E	8.31	9.30	5.28	22.90	76.9	40.9	R58.1	30.0	129.0	R4,918	R6,249	R5,685	5,636
1997E	10.44	11.33	5.70	27.47	79.2	52.0	R70.8	R33.8	156.7	R4,986	R6,254	R5,924	5,704
1998E	7.06	12.11	4.91	24.08	79.6	36.9	80.6	R32.2	149.6	R5,217	R6,655	R6,557	6,213
1999E	R4.09	10.51	R3.58	R18.18	80.3	R20.5	R64.1	R23.5	R108.1	R5,012	6,094	R6,565	R5,944
2000E	4.73	15.21	5.20	25.14	79.3	26.2	104.4	33.2	163.8	5,534	6,865	6,388	6,516

R=Revised. E=Estimate.

Notes: This table depicts all wells. Service wells, stratigraphic tests, and core tests are excluded. For 1949–1959, data represent wells completed in a given year. For 1960-1969, data are for well completion reports received by the American Petroleum Institute during the reporting year. For 1970 forward, the data represent wells completed in a given year. Totals may not equal sum of components due to independent rounding. Average depth may not equal average of components due to independent rounding.

SOURCE: "Table 4.4 Oil and Gas Exploratory and Development Wells, 1949–2000," in *Annual Energy Review 2000*, U.S. Department of Energy, Energy Information Administration, Washington, DC, 2001. EIA data sources include *World Oil* "Forecast Review" issue, Gulf Publishing Co.; *Quarterly Review of Drilling Statistics for the United States*, American Petroleum Institute, and well reports submitted to the API and Information Handling Services Energy Group, Inc.

development of offshore oil fields with onshore support in ANWR could mean significant human activity in the area during this century. In October 2002 Republicans in the U.S. Congress offered to designate much of the ANWR as wilderness in an effort to open the coastal plain of the refuge to oil drilling, but Democrats were opposed to that plan.

COAL

In 2000 the EIA estimated U.S. coal reserves at 502.7 billion short tons. (See Table 7.6.) About 43 percent of this, 213.8 billion short tons, is underground bituminous coal. Montana, Illinois, and Wyoming have the largest coal reserves.

In addition to untapped coal reserves, large stockpiles of coal are maintained by coal producers, distributors, and

major consumers (such as electric utility companies and industrial plants) to compensate for possible interruptions in supply. Although there is little seasonal change in demand for coal, supply can vary owing to factors such as coal miners' strikes and bad weather. According to the EIA's *Annual Energy Review 2000* (2001), coal stockpiles totaled 142.7 million short tons in 2000. Electric utilities held slightly more than 62 percent of this coal, and coal producers and distributors stocked another 24 percent.

URANIUM

The United States possesses enough uranium to fuel existing nuclear reactors for more than 40 years. In 2000 uranium reserves totaled 1.4 billion pounds of uranium oxide, mostly in Wyoming and New Mexico. (See Table 7.7.) Exploration for uranium has reflected that energy markets are moving away from nuclear energy. The number of exploratory and developmental holes drilled peaked at 104,350 in 1978 and declined to just 3,180 in 1999. (See Figure 7.5.)

INTERNATIONAL RESERVES

Crude Oil

Sixty-six percent of the estimated world crude oil reserves of approximately 1 trillion barrels (as of January 1, 2000) are located in the Middle East. (See Table 7.8. This graphic gives two different sets of estimates—one from Pennwell Publishing Company's *Oil & Gas Journal* and one from Gulf Publishing Company's *World Oil*.) Saudi Arabia, Iraq, United Arab Emirates, Kuwait, and Iran have the largest reserves. With an estimated 22 billion barrels of reserves, or only about 2 percent of the world's oil reserves total, the United States can no longer depend on its own oil reserves unless it drastically lowers consumption.

FIGURE 7.4

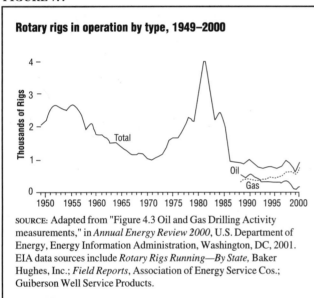

SOURCE: Adapted from "Figure 4.3 Oil and Gas Drilling Activity measurements," in *Annual Energy Review 2000*, U.S. Department of Energy, Energy Information Administration, Washington, DC, 2001. EIA data sources include *Rotary Rigs Running—By State*, Baker Hughes, Inc.; *Field Reports*, Association of Energy Service Cos.; Guiberson Well Service Products.

FIGURE 7.5

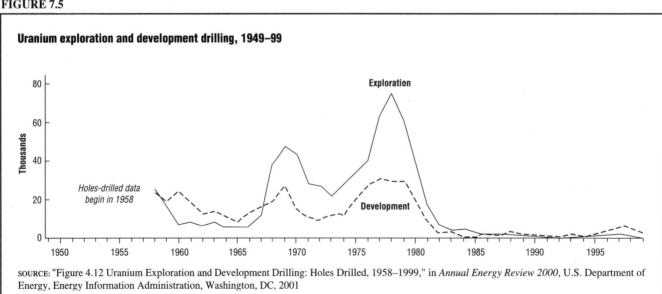

SOURCE: "Figure 4.12 Uranium Exploration and Development Drilling: Holes Drilled, 1958–1999," in *Annual Energy Review 2000*, U.S. Department of Energy, Energy Information Administration, Washington, DC, 2001

TABLE 7.4

Costs of oil and gas wells drilled, 1960–99

Year	Costs per well (thousand dollars)					Costs per foot (dollars)				
	Oil (nominal)	Gas (nominal)	Dry holes (nominal)	All (nominal)	All (real)[1]	Oil (nominal)	Gas (nominal)	Dry holes (nominal)	All (nominal)	All (real)
1960	52.2	102.7	44.0	54.9	247.6	13.22	18.57	10.56	13.01	58.63
1961	51.3	94.7	45.2	54.5	243.0	13.11	17.65	10.56	12.85	57.26
1962	54.2	97.1	50.8	58.6	257.9	13.41	18.10	11.20	13.31	58.53
1963	51.8	92.4	48.2	55.0	239.2	13.20	17.19	10.58	12.69	55.17
1964	50.6	104.8	48.5	55.8	239.2	13.12	18.57	10.64	12.86	55.10
1965	56.6	101.9	53.1	60.6	255.0	13.94	18.35	11.21	13.44	56.52
1966	62.2	133.8	56.9	68.4	279.6	15.04	21.75	12.34	14.95	61.12
1967	66.6	141.0	61.5	72.9	289.2	16.61	23.05	12.87	15.97	63.35
1968	79.1	148.5	66.2	81.5	309.7	18.63	24.05	12.88	16.83	63.99
1969	86.5	154.3	70.2	88.6	321.0	19.28	25.58	13.23	17.56	63.65
1970	86.7	160.7	80.9	94.9	326.5	19.29	26.75	15.21	18.84	64.83
1971	78.4	166.6	86.8	94.7	310.3	18.41	27.70	16.02	19.03	62.35
1972	93.5	157.8	94.9	106.4	334.5	20.77	27.78	17.28	20.76	65.24
1973	103.8	155.3	105.8	117.2	348.7	22.54	27.46	19.22	22.50	66.96
1974	110.2	189.2	141.7	138.7	378.8	27.82	34.11	26.76	28.93	79.00
1975	138.6	262.0	177.2	177.8	444.1	34.17	46.23	33.86	36.99	92.41
1976	151.1	270.4	190.3	191.6	453.0	37.35	49.78	36.94	40.46	95.65
1977	170.0	313.5	230.2	227.2	504.6	41.16	57.57	43.49	46.81	103.98
1978	208.0	374.2	281.7	280.0	580.4	49.72	68.37	52.55	56.63	117.42
1979	243.1	443.1	339.6	331.4	634.2	58.29	80.66	64.60	67.70	129.57
1980	272.1	536.4	376.5	367.7	644.6	66.36	95.16	73.70	77.02	135.03
1981	336.3	698.6	464.0	453.7	727.4	80.40	122.17	90.03	94.30	151.19
1982	347.4	864.3	515.4	514.4	776.4	86.34	146.20	104.09	108.73	164.12
1983	283.8	608.1	366.5	371.7	539.7	72.65	108.37	79.10	83.34	120.99
1984	262.1	489.8	329.2	326.5	457.0	66.32	88.80	67.18	71.90	100.64
1985	270.4	508.7	372.3	349.4	474.1	66.78	93.09	73.69	75.35	102.25
1986	284.9	522.9	389.2	364.6	484.1	68.35	93.02	76.53	76.88	102.08
1987	246.0	380.4	259.1	279.6	360.4	58.35	69.55	51.05	58.71	75.68
1988	279.4	460.3	366.4	354.7	442.2	62.28	84.65	66.96	70.23	87.56
1989	282.3	457.8	355.4	362.2	435.0	64.92	86.86	67.61	73.55	88.33
1990	321.8	471.3	367.5	383.6	443.4	69.17	90.73	67.49	76.07	87.93
1991	346.9	506.6	441.2	421.5	470.1	73.75	93.10	83.05	82.64	92.17
1992	362.3	426.1	357.6	382.6	416.6	69.50	72.83	67.82	70.27	76.51
1993	356.6	521.2	387.7	426.8	453.8	67.52	83.15	72.56	75.30	80.06
1994	409.5	535.1	491.5	483.2	503.3	70.57	81.90	86.60	79.49	82.79
1995	415.8	629.7	481.2	513.4	523.4	78.09	95.97	84.60	87.22	88.91
1996	341.0	616.0	541.0	496.1	496.1	70.60	98.67	95.74	88.92	88.92
1997	445.6	728.6	655.6	603.9	R592.4	90.48	117.55	115.09	107.83	R105.77
1998	566.0	815.6	973.2	769.1	R745.1	108.88	127.94	157.79	128.97	R124.95
1999	783.0	798.4	1115.5	856.1	817.2	156.45	138.42	182.99	152.02	145.10

[1]In chained (1996) dollars, calculated by using gross domestic product implicit price deflators.
R=Revised.
Notes: The information reported for 1965 and prior years is not strictly comparable to that in the more recent surveys. Average cost is the arithmetic mean and includes all costs for drilling and equipping wells and for surface-producing facilities. Wells drilled include exploratory and development wells; excludes service wells, stratigraphic tests, and core tests.

SOURCE: "Table 4.7 Costs of Oil and Gas Wells Drilled, 1960–1999," in *Annual Energy Review 2000*, U.S. Department of Energy, Energy Information Administration, Washington, DC, 2001. EIA data sources include *2000 Joint Association Survey on Drilling Costs*, American Petroleum Institute, Independent Petroleum Association of America, and Mid-Continent Oil and Gas Association.

Oil industry experts say that even though the world currently has sufficient supplies of oil, demand will rise in the future as economies such as those of China and India expand. This knowledge drives oil companies to seek new sources for oil. Companies from many nations have turned their attention to areas in the former Soviet Union, including the Caspian Sea, Azerbaijan, and Kazakhstan. Geologists believe that many billions of barrels of oil may be found there.

Many energy experts predict increased offshore exploration, including in deep water. Offshore production is expected to increase near Algeria, Nigeria, and Venezuela, as well as in the North Sea and offshore areas of West Africa, Brazil, Colombia, Mexico, and Canada. Deep-sea exploration is extremely expensive, but if oil prices rise, the oil industry may launch more deepwater explorations.

Natural Gas

Russia and the Middle East have approximately two-thirds of the world's estimated 5,150–5,211 trillion cubic feet (as of 2000) of natural gas reserves. (See Table 7.8.)

TABLE 7.5

Major U.S. energy companies' expenditures for oil and gas exploration and development by region, 1974–99
(billion dollars[1])

	United States			Foreign									
Year	Onshore	Offshore	Total	Canada	OECD[2] Europe	Eastern Europe and Former U.S.S.R.	Africa	Middle East	Other Eastern Hemisphere[3]	Other Western Hemisphere[4]	Total	Total
1974	NA	NA	8.7	NA	NA	—	NA	NA	NA	NA	3.8	12.5
1975	NA	NA	7.8	NA	NA	—	NA	NA	NA	NA	5.3	13.1
1976	NA	NA	9.5	NA	NA	—	NA	NA	NA	NA	5.2	14.7
1977	6.7	4.0	10.7	1.5	2.5	—	0.7	0.2	0.3	0.4	5.6	16.3
1978	7.5	4.3	11.8	1.6	2.6	—	0.8	0.3	0.4	0.6	6.4	18.2
1979	13.0	8.3	21.3	2.3	3.0	—	0.8	0.2	0.5	0.8	7.8	29.1
1980	16.8	9.4	26.2	3.1	4.3	—	1.4	0.2	0.8	1.0	11.0	37.2
1981	19.9	13.0	33.0	1.8	5.0	—	2.1	0.3	1.9	1.3	12.4	45.4
1982	27.2	11.9	39.1	1.9	6.3	—	2.1	0.4	2.4	1.1	14.2	53.3
1983	16.0	11.1	27.1	1.6	4.3	—	1.7	0.5	2.0	0.6	10.7	37.7
1984	32.1	16.0	48.1	5.4	5.5	—	3.4	0.5	2.0	0.5	17.3	65.3
1985	20.0	8.5	28.5	1.9	3.7	—	1.6	0.9	1.3	0.7	10.1	38.6
1986	12.5	4.9	17.4	1.1	3.2	—	1.1	0.3	1.2	0.6	7.5	24.9
1987	9.7	4.5	14.3	1.9	3.0	—	0.8	0.4	2.8	0.5	9.2	23.5
1988	12.9	8.1	21.0	5.4	4.3	—	0.8	0.4	1.4	0.7	13.0	34.1
1989	9.0	6.0	15.0	6.3	3.5	—	1.0	0.4	2.3	0.6	14.1	29.1
1990	10.2	4.9	15.1	1.8	6.6	—	1.4	0.6	2.4	0.7	13.6	28.7
1991	9.6	4.6	14.2	1.7	6.8	—	1.5	0.5	2.4	0.7	13.7	27.9
1992	7.3	3.0	10.3	1.1	6.8	—	1.4	0.6	2.4	0.6	12.9	23.2
1993	7.2	3.7	10.9	1.6	5.5	0.3	1.5	0.7	2.5	0.6	12.5	23.5
1994	7.8	4.8	12.6	1.8	4.4	0.3	1.4	0.4	2.8	0.7	11.9	24.5
1995	7.7	4.7	12.4	1.9	5.2	0.4	2.0	0.4	2.4	0.9	13.2	25.6
1996	7.9	6.7	14.6	1.6	5.6	0.5	2.8	0.5	4.1	1.6	16.6	31.3
1997	13.0	8.8	21.8	2.0	7.1	0.6	3.0	0.6	3.0	1.6	17.9	39.8
1998	13.5	11.0	24.4	4.8	8.6	1.3	3.1	0.9	3.9	3.7	26.4	50.8
1999	6.6	6.9	13.5	2.3	4.1	0.6	3.1	0.4	3.4	3.8	17.8	31.3

[1]Nominal dollars.
[2]Organization for Economic Cooporation and Development. Members of OCED as of December 31, 1999 were Australia, Austria, Belgium, Canada, Czech Republic, Denmark, Finland, France, Germany, Greece, Hungary, Iceland, Ireland, Italy, Japan, Luxembourg, Mexico, Netherlands, New Zealand, Norway, Poland, Portugal, South Korea, Spain, Sweden, Switzerland, Turkey, United Kingdom, and United States.
[3]This region includes areas that are eastward of the Greenwich prime meridian to 180° longitude and that are not included in other domestic or foreign classifications.
[4]This region includes areas that are westward of the Greenwich prime meridian to 180° longitude and that are not included in other domestic or foreign classifications.
— = Not applicable. NA=Not available.
Notes: Major U.S. Energy Companies are the top publicly-owned, U.S.-based crude oil and natural gas producers and petroleum refiners that form the Financial Reporting System (FRS).
Totals may not equal sum of components due to independent rounding.

SOURCE: "Table 4.9 Major U.S. Energy Companies' Expenditures for Oil and Gas Exploration and Development by Region, 1974–1999 (Billion Dollars)," in *Annual Energy Review 2000*, U.S. Department of Energy, Energy Information Administration, Washington, DC, 2001

Russia has more than twice as much natural gas in reserve as any other country, while Iran possesses (by far) the largest natural gas reserves in the Middle East. Large reserves (more than 100 trillion cubic feet) are also located in Qatar, United Arab Emirates, Saudi Arabia, the United States, Algeria, Venezuela, Nigeria, and Iraq.

Coal

In 1999 worldwide recoverable reserves of coal were estimated at about 1.1 trillion short tons. (See Table 7.9. Note that the data for the United States are for 1998, while those for other countries are 1996, the latest available.) The four countries with the most plentiful coal reserves are the United States (276 billion short tons), Russia (173 billion short tons), China (126 billion short tons), and Australia (100 billion short tons).

Uranium

The world's supply of uranium is much larger than the capacity for disposing it, should it be used as fuel. The countries with the largest known uranium reserves as of 1999, according to the Organization for Economic Cooperation and Development and the International Atomic Energy Agency, are Australia, Kazakhstan, Canada, South Africa, Namibia, Brazil, the Russian Federation, the United States, and Uzbekistan.

TABLE 7.6

Coal demonstrated reserve base, January 1, 2000

(billion short tons)

Region and State	Anthracite	Bituminous Coal Underground	Bituminous Coal Surface	Subbituminous Coal Underground	Subbituminous Coal Surface	Lignite Surface[1]	Total Underground	Total Surface	Total Total
Appalachian	**7.3**	**73.5**	**23.8**	**0.0**	**0.0**	**1.1**	**77.4**	**28.3**	**105.7**
Alabama	0.0	1.2	2.2	0.0	0.0	1.1	1.2	3.2	4.4
Kentucky, Eastern	0.0	1.8	9.7	0.0	0.0	0.0	1.8	9.7	11.5
Ohio	0.0	17.7	5.8	0.0	0.0	0.0	17.7	5.8	23.5
Pennsylvania	7.2	20.0	1.0	0.0	0.0	0.0	23.9	4.3	28.2
Virginia	0.1	1.3	0.6	0.0	0.0	0.0	1.4	0.6	2.0
West Virginia	0.0	30.3	4.2	0.0	0.0	0.0	30.3	4.2	34.5
Other[2]	0.0	1.2	0.4	0.0	0.0	0.0	1.2	0.4	1.5
Interior	**0.1**	**118.0**	**27.6**	**0.0**	**0.0**	**13.2**	**118.0**	**40.8**	**158.8**
Illinois	0.0	88.2	16.6	0.0	0.0	0.0	88.2	16.6	104.8
Indiana	0.0	8.8	0.9	0.0	0.0	0.0	8.8	0.9	9.8
Iowa	0.0	1.7	0.5	0.0	0.0	0.0	1.7	0.5	2.2
Kentucky, Western	0.0	16.1	3.7	0.0	0.0	0.0	16.1	3.7	19.8
Missouri	0.0	1.5	4.5	0.0	0.0	0.0	1.5	4.5	6.0
Oklahoma	0.0	1.2	0.3	0.0	0.0	0.0	1.2	0.3	1.6
Texas	0.0	0.0	0.0	0.0	0.0	12.7	0.0	12.7	12.7
Other[3]	0.1	0.3	1.1	0.0	0.0	0.5	0.4	1.6	2.0
Western	**(s)**	**22.4**	**2.3**	**121.3**	**62.3**	**29.6**	**143.8**	**94.3**	**238.1**
Alaska	0.0	0.6	0.1	4.8	0.6	(s)	5.4	0.7	6.1
Colorado	(s)	8.0	0.6	3.8	0.0	4.2	11.8	4.8	16.6
Montana	0.0	1.4	0.0	69.6	32.8	15.8	71.0	48.6	119.5
New Mexico	(s)	2.7	0.9	3.5	5.2	0.0	6.2	6.2	12.4
North Dakota	0.0	0.0	0.0	0.0	0.0	9.3	0.0	9.3	9.3
Utah	0.0	5.4	0.3	0.0	0.0	0.0	5.4	0.3	5.7
Washington	0.0	0.3	0.0	1.0	(s)	(s)	1.3	0.0	1.4
Wyoming	0.0	3.8	0.5	38.7	23.7	0.0	42.5	24.1	66.6
Other[4]	0.0	0.1	0.0	(s)	(s)	0.4	0.1	0.4	0.5
U.S. Total	**7.5**	**213.8**	**53.8**	**121.3**	**62.3**	**43.9**	**339.3**	**163.4**	**502.7**
States East of the Mississippi River	7.3	186.8	45.0	0.0	0.0	1.1	190.8	49.5	240.3
States West of the Mississippi River	0.1	27.0	8.7	121.3	62.3	42.8	148.5	113.9	262.4

[1]Lignite resources are not mined underground in the United States.
[2]Georgia, Maryland, North Carolina, and Tennessee.
[3]Arkansas, Kansas, Louisiana, and Michigan.
[4]Arizona, Idaho, Oregon, and South Dakota.
(s)=Less than 0.05 billion short tons.
Notes: Data represent known measured and indicated coal resources meeting minimum seam and depth criteria, in the ground as of January 1, 2000. These coal resources are not totally recoverable. Net recoverability ranges from 0 percent to more than 90 percent. Fifty-four percent of the demonstrated reserve base of coal in the United States is estimated to be recoverable. Totals may not equal sum of components due to independent rounding.

SOURCE: "Table 4.11 Coal Demonstrated Reserve Base, January 1, 2000 (Billion Short Tons)," in *Annual Energy Review 2000*, U.S. Department of Energy, Energy Information Administration, Washington, DC, 2001

TABLE 7.7

Uranium reserves and resources, 2000
(million pounds U₃O₈)

Resource category and state	Forward cost category (dollars per pound)[1]		
	$30 or less	$50 or less	$100 or less
Reserves[2]	**271**	**904**	**1,430**
New Mexico	84	341	567
Wyoming	110	370	591
Texas	7	24	40
Arizona, Colorado, Utah	41	115	160
Others[3]	29	54	73
Potential resources[4]			
Estimated additional resources	2,180	3,310	4,850
Speculative resources	1,310	2,230	3,480

[1]Forward costs are all operating and capital costs (in current dollars) yet to be incurred in the production of uranium from estimated resources. Excluded are previous expenditures (such as exploration and land acquisitions), taxes, profit, and the cost of money. Generally, forward costs are lower than market prices. Resource values in forward-cost categories are cumulative; that is, the quantity at each level of forward-cost includes all reserves/resources at the lower cost in that category.
[2]The Energy Information Administration category of uranium reserves is equivalent to the internationally reported category of Reasonably Assured Resources (RAR).
[3]California, Idaho, Nebraska, Nevada, North Dakota, Oregon, South Dakota, and Washington.
[4]Shown are the mean values for the distribution of estimates for each forward-cost category, rounded to the nearest million pounds U₃O₈.
Note: Data are at end of year.

SOURCE: "Table 4.13. Uranium Reserves and Resources, 2000," in *Annual Energy Review 2000*, U.S. Department of Energy, Energy Information Administration, Washington, DC, 2001

TABLE 7.8

World crude oil and natural gas reserves, January 1, 2000

Region and country	Crude oil (billion barrels)		Natural gas (trillion cubic feet)	
	Oil & Gas Journal	World Oil	Oil & Gas Journal	World Oil
North America	**55.1**	**55.6**	**261.3**	**261.3**
Canada	4.9	5.6	63.9	63.5
Mexico	28.4	28.3	30.1	30.4
United States	21.8	21.8	167.4	167.4
Central and South America	**89.5**	**69.2**	**222.7**	**227.9**
Argentina	2.8	2.6	24.2	24.3
Bolivia	0.1	0.2	4.3	5.5
Brazil	7.4	8.1	8.0	8.2
Colombia	2.6	2.3	6.9	6.6
Ecuador	2.1	3.0	3.7	3.9
Peru	0.4	4.1	9.0	8.8
Trinidad and Tobago	0.6	0.7	19.8	21.4
Venezuela	72.6	47.1	142.5	145.8
Other	1.0	1.0	4.2	3.5
Western Europe	**18.8**	**17.6**	**159.5**	**152.7**
Denmark	1.1	0.9	3.4	2.6
Germany	0.4	0.3	12.0	9.5
Italy	0.6	0.6	8.1	7.4
Netherlands	0.1	0.1	62.5	59.8
Norway	10.8	10.0	41.4	42.9
United Kingdom	5.2	5.0	26.7	26.8
Other	0.7	0.7	5.5	3.7
Eastern Europe and Former U.S.S.R.	**58.9**	**64.7**	**1,999.2**	**1,947.6**
Hungary	0.1	0.1	2.9	1.1
Kazakhstan	5.4	6.4	65.0	70.6
Romania	1.4	1.2	13.2	4.0
Russia	48.6	52.7	1,700.0	1,705.0
Other[1]	3.3	4.3	218.1	166.9

TABLE 7.8

World crude oil and natural gas reserves, January 1, 2000 [CONTINUED]

Region and country	Crude oil (billion barrels)		Natural gas (trillion cubic feet)	
	Oil & Gas Journal	World Oil	Oil & Gas Journal	World Oil
Middle East	**675.6**	**629.2**	**1,749.2**	**1,836.2**
Bahrain	0.1	NA	3.9	NA
Iran	89.7	93.1	812.3	790.0
Iraq	112.5	100.0	109.8	112.6
Kuwait	96.5	94.7	52.7	56.4
Oman	5.3	5.7	28.4	29.3
Qatar	3.7	5.4	300.0	394.0
Saudi Arabia	263.5	261.4	204.5	208.0
Syria	2.5	2.3	8.5	8.4
United Arab Emirates	97.8	63.8	212.0	209.0
Yemen	4.0	2.1	16.9	17.0
Other	(s)	0.5	0.3	11.5
Africa	**74.9**	**86.5**	**394.2**	**409.7**
Algeria	9.2	13.0	159.7	159.7
Angola	5.4	8.5	1.6	3.8
Cameroon	0.4	0.6	3.9	3.9
Congo	1.5	1.7	3.2	4.3
Egypt	2.9	3.8	35.2	42.5
Libya	29.5	29.5	46.4	46.4
Nigeria	22.5	24.5	124.0	126.0
Tunisia	0.3	0.3	2.8	2.8
Other	3.1	4.7	17.4	20.3
Far East and Oceania	**44.0**	**58.7**	**363.5**	**375.4**
Australia	2.9	2.9	44.6	44.6
Brunei	1.4	1.0	13.8	9.2
China	24.0	34.1	48.3	41.3
India	4.8	3.4	22.9	16.1
Indonesia	5.0	8.4	72.3	80.8
Malaysia	3.9	4.6	81.7	85.2
New Zealand	0.1	0.1	2.5	2.1
Pakistan	0.2	0.2	21.6	22.9
Papua New Guinea	0.3	0.8	5.4	17.3
Thailand	0.3	0.3	12.5	11.1
Other	1.1	2.9	37.9	44.7
World	**1,016.8**	**981.4**	**5,149.6**	**5,210.8**

[1]Albania, Azerbaijan, Belarus, Bulgaria, Czech Republic, Georgia, Kyrgyzstan, Lithuania, Poland, Slovakia, Tajikistan, Turkmenistan, Ukraine, Uzbekistan.
NA=Not available. (s)=Less than 0.05 billion barrels.
Notes: Data for Kuwait and Saudi Arabia include one-half of the reserves in the Neutral Zone between Kuwait and Saudi Arabia. All reserve figures except those for the former U.S.S.R. and natural gas reserves in Canada are proved reserves recoverable with present technology and prices at the time of estimation. Former U.S.S.R. and Canadian natural gas figures include proved, and some probable reserves. Totals may not equal sum of components due to independent rounding.

SOURCE: "Table 11.3. World Crude Oil and Natural Gas Reserves, January 1, 2000," in *Annual Energy Review 2000*, U.S. Department of Energy, Energy Information Administration, Washington, DC, 2001. EIA data sources include U.S. government reports as well as *Oil and Gas Journal*, PennWell Publishing Co., and *World Oil*, Gulf Publishing Co.

TABLE 7.9

World recoverable reserves of coal
(million short tons)

Region and country	Anthracite and bituminous coal	Subbituminous coal and lignite	Total
North America	R133,666	R152,948	R286,614
Canada	4,970	4,535	9,505
Greenland	0	202	202
Mexico	948	387	1,335
United States[1]	R127,748	R147,824	R275,572
Central and South America	**8,641**	**15,140**	**23,781**
Brazil	0	13,173	13,173
Chile	34	1,268	1,302
Colombia	7,020	420	7,439
Peru	1,058	110	1,168
Other	529	170	699
Western Europe	**29,022**	**70,636**	**99,658**
Germany	26,455	47,399	73,855
Greece	0	3,168	3,168
Serbia and Montenegro	71	18,087	18,157
Turkey	495	690	1,185
United Kingdom	1,102	551	1,653
Other	898	741	1,639
Eastern Europe and Former U.S.S.R.	**124,354**	**164,032**	**288,386**
Bulgaria	14	2,974	2,988
Czech Republic	2,880	3,929	6,809
Hungary	657	4,260	4,917
Kazakhstan	34,172	3,307	37,479
Poland	13,352	2,421	15,773
Romania	1	3,979	3,980
Russia	54,110	118,964	173,074
Ukraine	8,065	19,806	37,871
Uzbekistan	1,102	3,307	4,409
Other	0	1,085	1,085
Africa	**67,420**	**276**	**67,695**
Botswana	4,754	0	4,754
South Africa	60,994	0	60,994
Zimbabwe	809	0	809
Other	862	276	1,138
Middle East, Far East, and Oceania	**203,534**	**118,934**	**322,468**
Australia	52,139	47,510	99,649
China	68,564	57,651	126,215
India	80,174	2,205	82,379
Indonesia	849	4,905	5,754
Japan	865	0	865
Pakistan	0	3,228	3,228
Thailand	(s)	2,205	2,205
Other	942	1,231	2,174
World	R566,637	R521,965	R1,088,602

[1]U.S. data are more current than other data on this table. They represent recoverable reserves as of December 31, 1998; data for the other countries are as of December 31, 1996, the most recent period for which they are available.
R=Revised. (s)=Less than 0.5 million short tons.
Notes: World Energy Council data represent "Proved Recoverable Reserves," which are the tonnage within the Proved Amount in Place that can be recovered (extracted from the earth in raw form) under present and expected local economic conditions with existing, available technology. The EIA does not certify the international reserves data but reproduces the information as a matter of convenience for the reader. U.S. reserves represent estimated recoverable reserves from the Demonstrated Reserve Base which includes both measured and indicated tonnage. The U.S. term "measured" approximates the term "proved," used by the World Energy Council. The U.S. measured and indicated data have been combined and cannot be recaptured as "measured alone." Totals may not equal sum of components due to independent rounding.

SOURCE: "Table 11.12. World Recoverable Reserves of Coal," in *Annual Energy Review 2000*, U.S. Department of Energy, Energy Information Administration, Washington, DC, 2001. EIA data sources include U.S. government reports as well as *1998 Survey of Energy Resources*, World Energy Council.

CHAPTER 8
ELECTRICITY

Since 1879, when Thomas Edison flipped the first switch to light Menlo Park, New Jersey, the use of electrical power has become nearly universal in the United States.

ELECTRICITY DEFINED

Electricity is a form of energy resulting from the movement of charged particles, such as electrons (negatively charged subatomic particles) and protons (positively charged subatomic particles). Static electricity is caused by friction, when one material rubs against another and transfers charged particles. The zap you might feel and the spark you might see when you drag your feet along the carpet and then touch a metal doorknob is static electricity—electrons being transferred between you and the doorknob.

Electric current is the flow of electric charge; it is measured in amperes (amps). Electrical power is the rate at which energy is transferred by electric current. A watt is the standard measure of electrical power, named after the Scottish engineer James Watt. The term "wattage" refers to the amount of electrical power required to operate a particular appliance or device. A kilowatt is a unit of electrical power equal to 1,000 watts, while a kilowatt-hour is a unit of electrical work equal to that done by one kilowatt acting for one hour.

Electrical Capacity

The generating capacity of an electrical plant, measured in watts, indicates its ability to produce electrical power. A 1,000-kilowatt generator running at full capacity for one hour supplies 1,000 kilowatt-hours of power. That generator operating continuously for an entire year could produce 8.76 million kilowatt-hours of electricity (1,000 kilowatts x 24 hours/day x 365 days a year). However, no generator can operate at 100 percent capacity during an entire year because of downtime for routine maintenance, outages, and legal restrictions. On average, about one-fourth of the potential generating capacity of an electrical plant is not available at any given time.

Electricity demands vary daily and seasonally, so the continuous operation of electrical generators is not necessary. Utilities depend on steam, nuclear, and large hydroelectric plants to meet routine demand. Auxiliary gas, turbine, internal combustion, and smaller hydroelectric plants are normally used during short periods of high demand.

An Electric Power System

An electric power system has several components. Figure 8.1 illustrates a simple electric system. Generating units (power plants) produce electricity, transmission lines carry electricity over long distances, and distribution lines deliver the electricity to customers. Substations connect the pieces of the system together, while energy control centers coordinate the operation of all the components.

DOMESTIC ELECTRICITY USAGE

Domestic Production

A record 3.6 trillion kilowatt-hours of electricity was generated in 2000. Table 8.1 shows that electric utility retail sales (electricity use) in the United States has increased almost every year since 1950. Conventional steam plants, run by fossil fuels, were responsible for the production of about two-thirds of the electricity produced in 2000, or 2.1 trillion kilowatt-hours.

In the United States coal has been and continues to be the largest raw source for electricity production, accounting for more than half the electricity generated in 2000. (See Figure 8.2 and Figure 8.3.) Nuclear power was the second largest source of electricity, followed by natural gas, hydroelectric power, and petroleum. Extremely little electricity was generated by geothermal or other sources.

The Energy Information Administration (EIA) of the U.S. Department of Energy (DOE), in its *Annual Energy Review 2000* (published in 2001), noted that coal accounted

FIGURE 8.1

A simple electric system

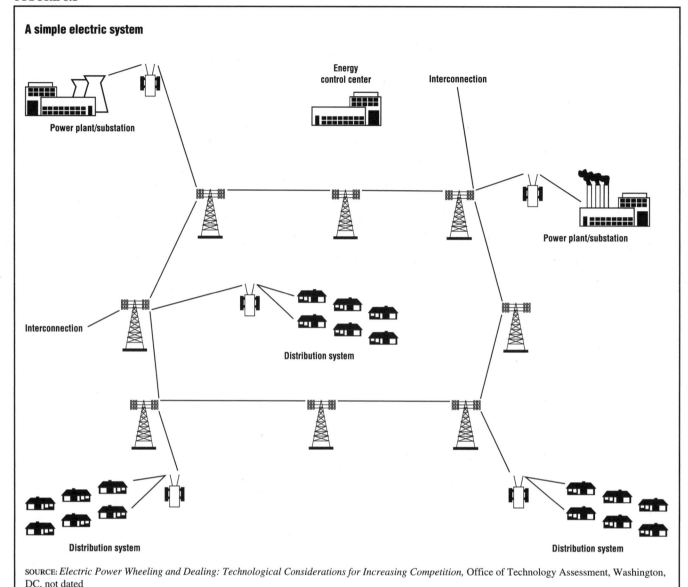

SOURCE: *Electric Power Wheeling and Dealing: Technological Considerations for Increasing Competition,* Office of Technology Assessment, Washington, DC, not dated

for the generation of 1.69 trillion kilowatt-hours of electricity in 2000, while natural gas contributed 290 billion kilowatt-hours and petroleum only 72 billion kilowatt-hours. Nuclear power accounted for 705 billion kilowatt-hours, and hydroelectric generation totaled 253 billion kilowatt-hours. All other renewable energy sources, including geothermal, wood, municipal waste, wind, and solar energy, produced 255 billion kilowatt-hours in 2000.

The structure of the electric power industry is evolving, moving away from traditional government-regulated electric utilities and toward an environment marked by less regulation and increased competition from nonutility power producers. In 2000, 21 percent of the total net generation of electricity came from nonutility producers, such as independent power producers and cogenerators. (See Figure 8.4.)

Domestic Consumption

From 1949 through 1990 the industrial sector was the largest consumer of electricity in the United States, but since then sales to the residential sector have generally been higher because of changing economic factors. (See Table 8.2.) In 2000 about 1.2 trillion kilowatt-hours went to residential use, 1.1 trillion kilowatt-hours to industrial customers, and 1.0 trillion kilowatt-hours to commercial users.

Consumption of electricity in general is growing because electricity has increasingly taken over the tasks formerly done with coal, natural gas, or human muscle: manufacturing steel, assembling cars, and milking cows. Electricity is being used extensively in rapidly growing technology fields, such as the computer industry. Residential and commercial use of electricity is increasing with new appliances, air conditioners, computers, and many other developing applications.

TABLE 8.1

Electricity overview, 1949–2000
(billion kilowatthours)

Year	Net generation			Imports[1]	Exports[1]	Losses and unaccounted for[2]	End use		
	Electric utilities	Nonutility power producers	Total				Electric utility retail sales[3]	Nonutility power producers[4]	Total
1949	291	NA	291	2	(s)	NA	255	NA	NA
1950	329	NA	329	2	(s)	NA	291	NA	NA
1951	371	NA	371	2	(s)	NA	330	NA	NA
1952	399	NA	399	3	(s)	NA	356	NA	NA
1953	443	NA	443	2	(s)	NA	396	NA	NA
1954	472	NA	472	3	(s)	NA	424	NA	NA
1955	547	NA	547	5	(s)	NA	497	NA	NA
1956	601	NA	601	5	1	NA	546	NA	NA
1957	632	NA	632	5	1	NA	576	NA	NA
1958	645	NA	645	4	1	NA	588	NA	NA
1959	710	NA	710	4	1	NA	647	NA	NA
1960	756	NA	756	5	1	NA	688	NA	NA
1961	794	NA	794	3	1	NA	722	NA	NA
1962	855	NA	855	2	2	NA	778	NA	NA
1963	917	NA	917	2	2	NA	833	NA	NA
1964	984	NA	984	6	4	NA	896	NA	NA
1965	1,055	NA	1,055	4	4	NA	954	NA	NA
1966	1,144	NA	1,144	4	3	NA	1,035	NA	NA
1967	1,214	NA	1,214	4	4	NA	1,099	NA	NA
1968	1,329	NA	1,329	4	4	NA	1,203	NA	NA
1969	1,442	NA	1,442	5	4	NA	1,314	NA	NA
1970	1,532	NA	1,532	6	4	NA	1,392	NA	NA
1971	1,613	NA	1,613	7	4	NA	1,470	NA	NA
1972	1,750	NA	1,750	10	3	NA	1,595	NA	NA
1973	1,861	NA	1,861	17	3	NA	1,713	NA	NA
1974	1,867	NA	1,867	15	3	NA	1,706	NA	NA
1975	1,918	NA	1,918	11	5	NA	1,747	NA	NA
1976	2,038	NA	2,038	11	2	NA	1,855	NA	NA
1977	2,124	NA	2,124	20	3	NA	1,948	NA	NA
1978	2,206	NA	2,206	21	1	NA	2,018	NA	NA
1979	2,247	NA	2,247	23	2	NA	2,071	NA	NA
1980	2,286	NA	2,286	25	4	NA	2,094	NA	NA
1981	2,295	NA	2,295	36	3	NA	2,147	NA	NA
1982	2,241	NA	2,241	33	4	NA	2,086	NA	NA
1983	2,310	NA	2,310	39	3	NA	2,151	NA	NA
1984	2,416	NA	2,416	42	3	NA	2,286	NA	NA
1985	2,470	NA	2,470	46	5	NA	2,324	NA	NA
1986	2,487	NA	2,487	41	5	NA	2,369	NA	NA
1987	2,572	NA	2,572	52	6	NA	2,457	NA	NA
1988	2,704	NA	2,704	39	7	NA	2,578	NA	NA
1989	2,784	[5]188	2,972	26	15	236	2,647	[5]100	2,747
1990	2,808	[5]217	3,025	18	16	210	2,713	[5]104	2,817
1991	2,825	[5]246	3,071	22	2	218	2,762	[5]111	2,873
1992	2,797	286	3,083	28	3	224	2,763	122	2,885
1993	2,883	314	3,197	31	4	236	2,861	127	2,988
1994	2,911	343	3,254	47	2	223	2,935	141	3,075
1995	2,995	363	3,358	43	4	235	3,013	149	3,162
1996	3,077	370	3,447	43	3	R237	R3,101	149	R3,250
1997	3,123	372	3,494	43	9	R234	R3,146	149	R3,295
1998	3,212	406	3,618	40	13	R220	R3,264	160	R3,424
1999	R3,174	R532	R3,706	43	14	234	R3,312	189	3,501
2000P	3,010	E782	3,792	50	15	221	3,398	F208	3,607

[1]Electricity transmitted across U.S. borders with Canada and Mexico.
[2]Energy losses that occur between the point of generation and delivery to the customer, and data collection frame differences and nonsampling error.
[3]Includes nonutility sales of electricity to utilities for distribution to end users. Beginning in 1996, also includes sales to ultimate consumers by power marketers.
[4]Nonutility facility use of onsite net electricity generation, and nonutility sales of electricity to end users.
[5]Data for 1989-1991 were collected for facilities with capacities of 5 megawatts or more. In 1992, the threshold was lowered to include facilities with capacities of 1 megawatt or more. Estimates of the 1-to-5 megawatt range for 1989-1991 were derived from historical data. The estimation did not include retirements that occurred prior to 1992 and included only the capacity of facilities that came on line before 1992.
R=Revised. P=Preliminary. E=Estimate. F=Forecast. NA=Not available. (s)=Less than 0.5 billion kilowatthours.
Totals may not equal sum of components due to independent rounding.

SOURCE: "Table 8.1. Electricity Overview, 1949–2000," in *Annual Energy Review 2000*, U.S. Department of Energy, Energy Information Administration, Washington, DC, 2001

FIGURE 8.2

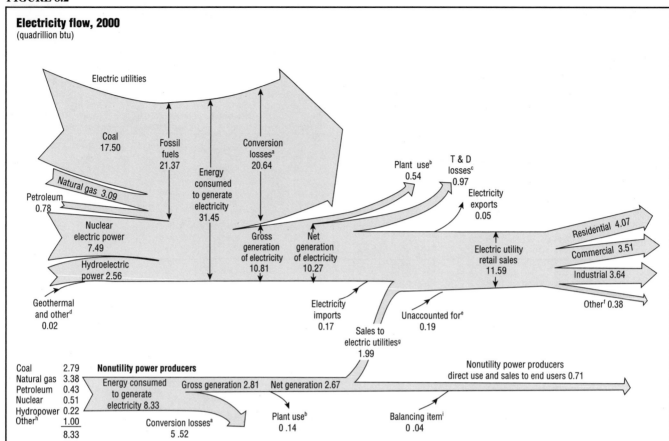

Electricity flow, 2000
(quadrillion btu)

[a]Approximately two-thirds of all energy used to generate electricity. Electrical system energy losses are estimated as the difference between total energy consumed to generate electricity and the total energy content of electricity consumed by end users. Most of these losses occur at steam-electric power plants (conventional and nuclear) in the conversion of heat energy into mechanical energy to turn electric generators. This loss is a thermodynamically necessary feature of the steam-electric cycle. Part of the energy input-to-output losses are a result of imputing fossil energy equivalent inputs for hydroelectric and other energy sources, since there is no generally accepted practice for measuring these thermal conversion rates. In addition to conversion losses, other losses include power plant use of electricity, transmission and distribution of electricity from power plants to end-use consumers (also called "line-losses"), and unaccounted-for electricity. Total losses are allocated to the end-use sectors in proportion to each sector's share of total electricity sales. Overall approximately 67 percent of total energy input is lost in conversion: of electricity generated, approximately 5 percent is lost in plant use and 9 percent is lost in transmission and distribution. Calculated electrical energy system losses may be less than actual losses, because primary consumption does not include the energy equivalent of electricity imports from Canada and Mexico, although they are included in electricity end use.
[b]The electric energy used in the operation of power plants, estimated as 5 percent of gross generation. See also footnote[a].
[c]Transmission and distribution losses are estimated as 9 percent of gross generation of electricity. See also footnote[a].
[d]Wood, waste, wind, and solar energy used to generate electricity.
[e]Balancing item to adjust for data collection frame differences and nonsampling error.
[f]Public street and highway lighting, other sales to public authorities, sales to railroads and railways, and interdepartmental sales.
[g]Sales, interchanges, and exchanges of electric energy with utilities.
[h]Geothermal, wood, waste, wind, and solar energy used to generate electricity.
[i]Transmission and distribution losses and unaccounted for.
Note: Totals may not equal sum of components due to independent rounding.

SOURCE: "Diagram 5. Electricity flow, 2000," in *Annual Energy Review 2000*, U.S. Department of Energy, Energy Information Administration, Washington, DC, 2001.

THE ELECTRIC BILL

The price paid by a consumer for electricity includes the cost of converting energy into electricity from its original form, such as coal, as well as the cost of delivering it. In 1997, according to the EIA's *Annual Energy Review 2000* (2001), consumers paid an average of $20.15 per million Btu for the electric power delivered to their residences, compared to only $4.62 per million Btu for natural gas and $9.73 per million Btu for motor gasoline.

The unit cost of electricity is high because of the amounts of energy expended in creating the electricity and moving it to the point of use. In 1999, for example, about 31.5 quadrillion Btu of energy were consumed by electric utilities to generate electricity in the United States, but 10.3 quadrillion Btu was the net generation, after accounting for energy used by the power plants themselves. (See Figure 8.2.) Most of the remaining 21.2 quadrillion Btu was lost during the energy conversion process. Additionally, about 1 quadrillion Btu is lost during the transmission and distribution process (T & D losses). In the end, for every three units of energy that are converted to create electricity, slightly less than one unit actually reaches the end user.

FIGURE 8.3

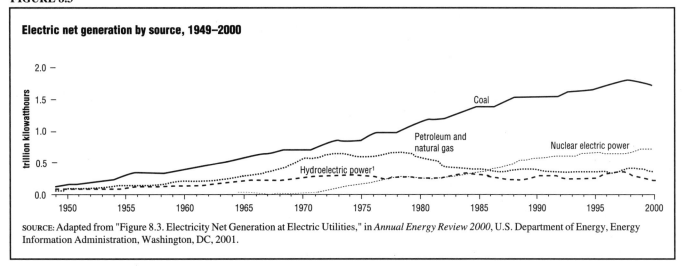

Electric net generation by source, 1949–2000

SOURCE: Adapted from "Figure 8.3. Electricity Net Generation at Electric Utilities," in *Annual Energy Review 2000*, U.S. Department of Energy, Energy Information Administration, Washington, DC, 2001.

Between 1935 and 1970, the price of electricity declined, but it began to increase during the 1970s because of the OPEC oil embargo. (See Figure 8.5.) Since 1985 the price of electricity has been dropping again because of the decline in energy resource prices. Prices vary depending upon the location. As Figure 8.6 shows, in 2000 electricity was most expensive in the New England states, New York, New Jersey, Maryland, Washington, D.C., Michigan, Arizona, California, Alaska, and Hawaii. According to the EIA's *Electric Power Annual 2000, Volume 1* (August 2001), the average price of electricity sold to the residential sector was 8.2 cents per kilowatt-hour in 2000, while the commercial sector paid 7.2 cents per kilowatt-hour. Industrial users paid less per kilowatt-hour, 4.5 cents in 2000, because the huge amounts of electricity they use allow them to receive volume discounts. The average price for all sectors across the United States in 2000 was 6.7 cents.

DEREGULATION OF ELECTRIC UTILITIES

Regulated for decades as "natural monopolies," much like the railroad and telecommunications industries, electric utilities are in the midst of a radical shift toward unregulated markets and increased competition. In 1978 Congress passed the Public Utilities Regulatory Policy Act (PURPA; PL 95-617), which required that utilities buy electricity from private companies when that would be a lower-cost alternative to building their own power plants. The Energy Policy Act of 1992 (PL 102-486) gave other generators greater access to the market, resulting in a flurry of activity in state and federal legislatures as a host of interest groups debated regulatory, economic, energy, and environmental policies. State public utility commissions are conducting proceedings and designing rules related to competition in the electric utility industry. As of October 2002, 17 states plus the District of Columbia were actively engaged in restructuring activities. (See Figure 8.7.) In six states, restructuring was delayed, and

FIGURE 8.4

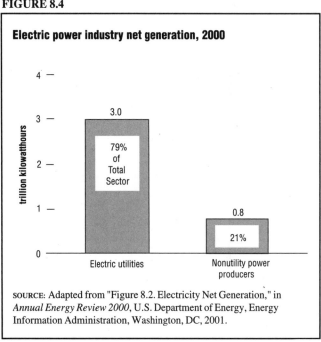

Electric power industry net generation, 2000

SOURCE: Adapted from "Figure 8.2. Electricity Net Generation," in *Annual Energy Review 2000*, U.S. Department of Energy, Energy Information Administration, Washington, DC, 2001.

in one state (California) restructuring was suspended. Restructuring was not active in 26 states.

In 1998 in California, the first state to offer deregulated options to consumers, a cautious public was bombarded with aggressive sales pitches and confusing claims by many electricity marketers. Most residents, however, remained with their original utility providers. During peak periods in 2000 and 2001, many parts of California experienced severe shortages of electricity, necessitating rolling blackouts. This occurred because several of California's largest utility companies were facing bankruptcy. Therefore, they could not afford to buy wholesale electricity from other producers to meet all of their customers' demands. Californians also experienced extremely high electricity prices. Both situations caused the public to question whether deregulation was going to work at all.

TABLE 8.2

Electricity end use, 1949–2000

(billion kilowatthours)

| Year | Electric utility retail sales[1] | | | | | Nonutility power producers | | | Total |
	Residential	Commercial	Industrial	Other[2]	Total	Direct use[3]	Sales to end users	Total	
1949	67	45	123	20	255	NA	NA	NA	NA
1950	72	51	146	22	291	NA	NA	NA	NA
1951	83	57	166	24	330	NA	NA	NA	NA
1952	94	62	176	24	356	NA	NA	NA	NA
1953	104	67	199	26	396	NA	NA	NA	NA
1954	116	72	208	27	424	NA	NA	NA	NA
1955	128	79	260	29	497	NA	NA	NA	NA
1956	143	87	286	30	546	NA	NA	NA	NA
1957	157	94	294	31	576	NA	NA	NA	NA
1958	169	100	287	32	588	NA	NA	NA	NA
1959	185	112	315	36	647	NA	NA	NA	NA
1960	201	131	324	32	688	NA	NA	NA	NA
1961	214	138	337	32	722	NA	NA	NA	NA
1962	233	153	360	32	778	NA	NA	NA	NA
1963	251	171	377	34	833	NA	NA	NA	NA
1964	272	187	405	32	896	NA	NA	NA	NA
1965	291	200	429	34	954	NA	NA	NA	NA
1966	317	218	464	37	1,035	NA	NA	NA	NA
1967	340	234	485	40	1,099	NA	NA	NA	NA
1968	382	258	521	42	1,203	NA	NA	NA	NA
1969	427	282	559	46	1,314	NA	NA	NA	NA
1970	466	307	571	48	1,392	NA	NA	NA	NA
1971	500	329	589	51	1,470	NA	NA	NA	NA
1972	539	359	641	56	1,595	NA	NA	NA	NA
1973	579	388	686	59	1,713	NA	NA	NA	NA
1974	578	385	685	58	1,706	NA	NA	NA	NA
1975	588	403	688	68	1,747	NA	NA	NA	NA
1976	606	425	754	70	1,855	NA	NA	NA	NA
1977	645	447	786	71	1,948	NA	NA	NA	NA
1978	674	461	809	73	2,018	NA	NA	NA	NA
1979	683	473	842	73	2,071	NA	NA	NA	NA
1980	717	488	815	74	2,094	NA	NA	NA	NA
1981	722	514	826	85	2,147	NA	NA	NA	NA
1982	730	526	745	86	2,086	NA	NA	NA	NA
1983	751	544	776	80	2,151	NA	NA	NA	NA
1984	780	583	838	85	2,286	NA	NA	NA	NA
1985	794	606	837	87	2,324	NA	NA	NA	NA
1986	819	631	831	89	2,369	NA	NA	NA	NA
1987	850	660	858	88	2,457	NA	NA	NA	NA
1988	893	699	896	90	2,578	NA	NA	NA	NA
1989	906	726	926	90	2,647	[4]83	[4]18	100	2,747
1990	924	751	946	92	2,713	[4]84	[4]20	104	2,817
1991	955	766	947	94	2,762	[4]100	[4]11	111	2,873
1992	936	761	973	93	2,763	111	11	122	2,885
1993	995	795	977	95	2,861	111	16	127	2,988
1994	1,008	820	1,008	98	2,935	123	18	141	3,075
1995	1,043	863	1,013	95	3,013	134	16	149	3,162
1996	R1,083	887	R1,034	98	R3,101	135	14	149	R3,250
1997	1,076	R929	R1,038	103	R3,146	131	18	149	R3,295
1998	R1,130	R979	R1,051	104	R3,264	134	26	160	R3,424
1999	R1,145	R1,002	R1,058	R107	R3,312	147	42	189	3,501
2000P	1,192	1,028	1,068	110	3,398	NA	NA	F208	3,607

[1]Includes nonutility sales of electricity to utilities for distribution to end users. Beginning in 1996, also includes sales to ultimate consumers by power marketers.
[2]Public street and highway lighting, other sales to public authorities, sales to railroads and railways, and interdepartmental sales.
[3]Nonutility facility use of onsite net electricity generation.
[4]Data for 1989-1991 were collected for facilities with capacities of 5 megawatts or more. In 1992, the threshold was lowered to include facilities with capacities of 1 megawatt or more. Estimates of the 1-to-5 megawatt range for 1989-1991 were derived from historical data. The estimation did not include retirements that occurred prior to 1992 and included only the capacity of facilities that came on line before 1992.
R=Revised. P=Preliminary. F=Forecast. NA=Not available.
Totals may not equal sum of components due to independent rounding.

SOURCE: "Table 8.12. Electricity End Use, 1949–2000," in *Annual Energy Review 2000*, U.S. Department of Energy, Energy Information Administration, Washington, DC, 2001

The utilities blamed the slow process of deregulation for the problems, saying that it had forced them into paying high market prices for wholesale electricity even though they were not allowed to pass those costs on to their customers. The result was that the utilities had no money left with which to procure more electricity. The California state government eventually had to step in and help the utilities find affordable wholesale electricity.

On April 15, 1999, the Clinton administration submitted to Congress its proposed Comprehensive Electricity Competition Act (CECA), which would further guide and empower the states to increase competition in the electricity market and ultimately reduce consumer prices. The act would also have provisions to stimulate research and investment in renewable energy and efficiency, give consumers choice and protection, and reduce emissions. As of late 2002, the Senate Committee on Energy and Natural Resources had held hearings on the CECA, but the bill was still pending ratification by Congress.

INTERNATIONAL ELECTRICITY USAGE

World Production

In 1999 more than 14 trillion kilowatt-hours of electricity were generated around the world. (See Table 8.3.) North America accounted for 32 percent; the Far East and Oceania, 27 percent; western Europe, 20 percent; and eastern Europe and the former USSR, 11 percent. In 1999 fossil fuels (coal, gas, and oil) accounted for 63 percent of all electricity generated; hydroelectric, 19 percent; nuclear, 17 percent; and geothermal and other sources, less than 1 percent of the world's total net electricity production.

World Consumption

World total electricity consumption continues to increase, rising from 10.8 trillion kilowatt-hours in 1991 to 13.7 trillion kilowatt-hours in 2000. (See Table 8.4.) North America accounted for 31 percent of the world's total consumption in 2000; Asia and Oceania, 27 percent; western Europe, 20 percent; and eastern Europe and the former USSR, 10 percent.

FUTURE TRENDS IN THE U.S. ELECTRIC INDUSTRY

The EIA's *Annual Energy Outlook 2002* (2001) predicts that for the next 20 years, residential U.S. electricity consumption will grow at a rate of 1.8 percent annually. This compares with 7 percent growth per year during the 1960s, 1.5 percent growth in the 1970s, and 1.0 percent in the 1980s. Several factors led to this decreased growth of electricity consumption, including increased market saturation of electric appliances and improvements in efficiency. From 2000 to 2020, residential electricity demand is projected to grow by 1.7 percent per year because of an increase in the number of U.S. households. Commercial

FIGURE 8.5

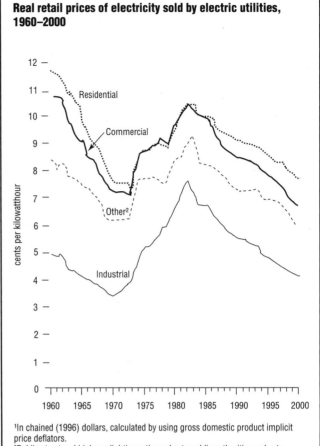

Real retail prices of electricity sold by electric utilities, 1960–2000

[1]In chained (1996) dollars, calculated by using gross domestic product implicit price deflators.
[2]Public street and highway lighting, other sales to public authorities, sales to railroads and railways, and interdepartmental sales.

SOURCE: Adapted from "Figure 8.15. Retail Prices of Electricity Sold by Electric Utilities, 1960–2000," in *Annual Energy Review 2000*, U.S. Department of Energy, Energy Information Administration, Washington, DC, 2001.

demand is expected to grow by 2.3 percent per year because of growth in commercial floor space, while industrial demand will likely increase by 1.4 percent per year as industrial output rises.

Historically, the demand for electricity in the United States has been related to economic growth. This relationship will continue, but electricity use is expected to grow more slowly than the gross domestic product (GDP), a measure of economic growth. Figure 8.8 shows how electricity sales are related more to economic growth (the GDP) than to population growth. Although the lines on the graph decrease from about 2010 to 2020, the Y-axis stands for growth. Therefore, a descending line means that growth is slowing; however, growth is still taking place. Also, the phrase "5-year moving average" means that each point on the graph is an average of that year and the previous four years' data. This type of averaging is often done to get a better idea of what the real long-term averages would be, without heavy influences from cyclical influences on data, like business or weather cycles.

FIGURE 8.6

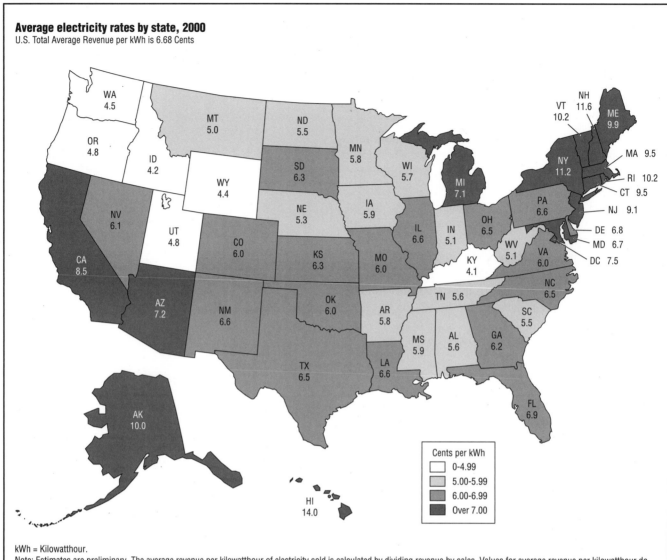

Average electricity rates by state, 2000
U.S. Total Average Revenue per kWh is 6.68 Cents

Cents per kWh
- ☐ 0-4.99
- ☐ 5.00-5.99
- ☐ 6.00-6.99
- ☐ Over 7.00

kWh = Kilowatthour.
Note: Estimates are preliminary. The average revenue per kilowatthour of electricity sold is calculated by dividing revenue by sales. Values for average revenue per kilowatthour do not account for all energy service providers. Consequently, the growth in sales is underestimated (in particular for the commercial and industrial sectors). This, in turn, may affect the rates of associated revenue to sales of electricity.

SOURCE: "Figure 12. Estimated Average Revenue per Kilowatthour for All Sectors at Electric Utilities by State, 2000," in *Electric Power Annual 2000,* vol. 1, Energy Information Administration, Washington, DC, August, 2001.

The issue of electric growth is important and carries financial risks for electric companies. If the industry underestimates future needs for electricity, it could mean power shortages or losses. However, excessive projections of the nation's needs could mean billions of dollars spent on unneeded equipment.

The EIA estimates that the United States will need 355 gigawatts of new generating capacity by 2020 to meet growing demand for electricity and to replace retiring units. Because of plant retirements, nuclear energy is expected to lose 10 gigawatts (10 percent) of its current capacity, and fossil-fueled energy is expected to lose 37 gigawatts (7 percent) of its capacity by 2020. The EIA projects that 312 gigawatts (88 percent) of the new generating capacity needed will be fueled by new natural gas capacity, 31 gigawatts

(9 percent) by new coal capacity, and the remaining 3 percent by new renewable technology capacity (primarily wind, geothermal, and municipal solid waste units).

Electricity prices are expected to decrease by an average of 0.3 percent per year from 2000 to 2020 as a result of competition among electricity suppliers. Therefore, average electricity prices across sectors are projected to decline from 6.2 cents per kilowatt-hour in 2000 to 5.8 cents in 2020.

Growing concerns about acid rain and global warming could result in tightened environmental emission standards, which may have an impact on electrical utility expansion decisions, prices, and supply. Continued advances in solar and wind turbine technology could make renewable sources of electrical power more economical in

FIGURE 8.7

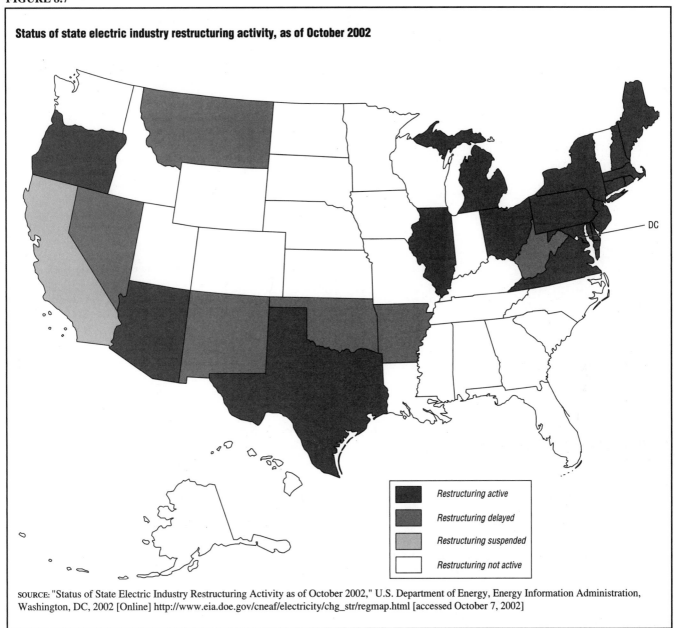

Status of state electric industry restructuring activity, as of October 2002

Restructuring active

Restructuring delayed

Restructuring suspended

Restructuring not active

SOURCE: "Status of State Electric Industry Restructuring Activity as of October 2002," U.S. Department of Energy, Energy Information Administration, Washington, DC, 2002 [Online] http://www.eia.doe.gov/cneaf/electricity/chg_str/regmap.html [accessed October 7, 2002]

the future. Some energy experts and environmentalists claim that increased efficiency and conservation efforts are the most sensible alternatives to new construction or to the burning of more fossil fuels in existing plants.

TABLE 8.3

World net generation of electricity by type, 1980, 1998, and 1999

(billion kilowatthours)

Region and country	Fossil fuel 1980	Fossil fuel 1998	Fossil fuel 1999[P]	Nuclear electric power 1980	Nuclear electric power 1998	Nuclear electric power 1999[P]	Hydroelectric power[1] 1980	Hydroelectric power[1] 1998	Hydroelectric power[1] 1999[P]	Total[2] 1980	Total[2] 1998	Total[2] 1999[P]
North America	**1,880.3**	**R2,833.8**	**2,863.1**	**287.0**	**R750.2**	**807.6**	**546.9**	**R671.8**	**685.9**	**R2,718.4**	**R4,342.5**	**4,456.6**
Canada	79.8	R148.7	149.6	35.9	R 67.7	69.8	251.0	R328.6	340.3	367.9	R551.1	567.2
Mexico	46.0	R134.2	135.3	0.0	8.8	9.5	16.7	R24.4	32.4	63.6	R172.8	182.5
United States	R1,753.8	R2,550.0	2,577.4	251.1	673.7	728.3	279.2	318.9	313.2	R2,286.4	3,617.9	3,706.1
Other	0.5	R0.8	0.8	0.0	0.0	0.0	0.0	0.0	0.0	0.5	R0.8	0.8
Central and South America	**99.8**	**R177.3**	**197.7**	**2.2**	**10.3**	**10.5**	**201.5**	**R521.4**	**528.5**	**R308.2**	**R723.4**	**752.2**
Argentina	22.2	R37.5	46.5	2.2	7.1	6.7	17.3	35.8	23.7	R41.5	R80.6	77.1
Brazil	7.5	15.6	17.8	0.0	3.1	3.8	128.4	288.5	305.9	138.3	316.9	337.4
Paraguay	(s)	0.1	(s)	0.0	0.0	0.0	0.7	50.3	51.4	0.8	50.4	51.6
Venezuela	17.6	R21.6	26.1	0.0	0.0	0.0	14.4	52.5	55.1	32.0	R74.0	81.2
Other	52.4	R102.5	107.3	0.0	0.0	0.0	R40.6	R94.4	92.3	R95.7	R201.4	204.9
Western Europe	**1,180.1**	**R1,327.1**	**1,346.7**	**219.2**	**R841.0**	**850.2**	**431.7**	**R519.8**	**523.1**	**1,844.5**	**R2,746.6**	**2,785.0**
Belgium	38.3	33.4	31.9	11.9	43.9	46.6	0.3	0.4	0.3	50.8	78.7	79.8
Finland	22.0	R31.2	31.7	6.6	20.8	21.8	10.1	R14.9	12.7	38.7	R76.2	75.8
France	118.0	R52.9	48.2	63.4	R368.6	375.1	68.3	R61.4	71.6	250.8	R485.7	497.3
Germany	390.3	R347.0	336.3	55.6	R153.6	161.0	18.8	R17.0	19.1	469.9	R530.2	531.4
Italy	125.5	194.8	195.9	2.1	0.0	0.0	45.0	40.8	44.8	176.4	241.7	247.7
Netherlands	58.0	81.2	77.0	3.9	3.6	3.6	0.0	0.1	0.1	62.9	89.2	85.3
Norway	0.1	0.7	0.8	0.0	0.0	0.0	82.7	R114.2	120.0	82.9	R115.2	121.1
Spain	74.5	R93.2	114.1	5.2	56.0	55.9	29.2	R33.7	23.9	109.2	R186.4	197.7
Sweden	10.1	R9.4	8.1	25.3	R69.9	66.6	58.1	R73.6	69.3	94.3	R156.1	146.6
Switzerland	0.9	2.3	2.3	12.9	24.5	23.7	32.5	R33.1	39.5	46.4	61.1	66.8
Turkey	12.0	64.6	77.0	0.0	0.0	0.0	11.2	41.8	34.3	23.3	106.7	111.5
United Kingdom	228.9	R235.3	237.8	32.3	R95.1	91.5	3.9	R5.2	5.3	265.1	R341.9	342.8
Other	R101.4	R181.1	185.6	R0.0	R 5.0	4.5	R71.7	R83.6	82.3	R173.8	R277.6	281.3
Eastern Europe and Former U.S.S.R.	**1,309.3**	**R1,010.0**	**999.5**	**83.2**	**239.3**	**245.7**	**211.3**	**R261.3**	**261.7**	**1,604.1**	**R1,512.3**	**1,508.6**
Czech Republic	—	R46.6	45.3	—	12.5	12.7	—	R1.7	1.6	—	61.5	60.7
Kazakhstan	—	40.4	38.6	—	0.1	0.1	—	6.1	5.6	—	46.6	44.4
Poland	111.1	R130.2	129.6	0.0	0.0	0.0	3.2	4.3	4.2	114.7	135.0	134.4
Romania	51.4	R27.5	26.5	0.0	4.9	4.8	12.5	R18.7	17.7	63.9	R51.1	49.0
Russia	—	R530.1	529.2	—	98.3	110.9	—	R157.9	157.9	—	R786.3	798.1
Ukraine	—	R83.2	76.0	—	70.6	67.4	—	R15.8	15.2	—	R163.2	157.8
Other	1,146.8	R152.0	154.4	83.2	**52.8**	49.8	195.5	56.9	59.4	1,425.6	R268.5	264.3

TABLE 8.3

World net generation of electricity by type, 1980, 1998, and 1999 [CONTINUED]

(billion kilowatthours)

Region and country	Fossil fuel			Nuclear electric power			Hydroelectric power[1]			Total[2]		
	1980	1998	1999P	1980	1998	1999P	1980	1998	1999P	1980	1998	1999P
Middle East	**82.8**	**R379.8**	**401.7**	**0.0**	**0.0**	**0.0**	**9.6**	**R15.8**	**16.0**	**92.4**	**R395.6**	**417.7**
Iran	15.7	R90.6	96.0	0.0	0.0	0.0	5.6	R6.9	7.1	21.3	R97.6	103.1
Saudi Arabia	20.5	R116.5	120.0	0.0	0.0	0.0	0.0	0.0	0.0	20.5	R116.5	120.0
Other	46.6	R172.7	185.7	0.0	0.0	0.0	4.1	R8.9	8.9	50.7	R181.6	194.6
Africa	**129.1**	**R311.0**	**315.3**	**0.0**	**13.6**	**12.8**	**60.6**	**63.1**	**65.6**	**189.7**	**R388.0**	**394.1**
Egypt	8.6	R47.1	49.5	0.0	0.0	0.0	9.7	R12.1	15.1	18.3R	59.2	64.7
South Africa	92.1	R176.5	173.3	0.0	13.6	12.8	1.0	1.6	0.7	93.1	R191.7	186.9
Other	28.4	R87.4	92.5	0.0	0.0	0.0	49.9	R49.4	49.7	78.4	R137.1	142.5
Far East and Oceania	**907.7**	**R2,544.9**	**2,689.8**	**92.7**	**R460.8**	**469.1**	**275.2**	**R513.7**	**534.1**	**1,280.5**	**R3,565.4**	**3,742.2**
Australia	74.5	R167.8	172.4	0.0	0.0	0.0	12.8	15.6	16.0	87.7	R186.6	191.7
China	227.9	R880.2	936.5	0.0	13.5	14.1	57.6	202.9	222.8	285.5	R1,096.5	1,173.4
India	69.7	R337.2	361.0	3.0	10.6	11.5	46.5	R75.5	80.8	119.3	R424.3	454.6
Indonesia	10.6	R58.6	63.2	0.0	0.0	0.0	3.0	R10.5	11.5	13.5	R72.7	78.7
Japan	381.6	R571.3	599.9	78.6	R315.7	308.7	87.8	R91.6	85.0	549.1	R1,002.4	1,018.3
South Korea	29.8	R132.9	148.2	3.3	85.2	97.9	1.5	R4.1	4.1	34.6	R222.3	250.3
Taiwan	31.3	R91.7	93.9	7.8	R35.4	36.9	2.9	R9.9	8.8	42.0	R137.0	139.7
Thailand	12.3	R76.8	81.5	0.0	0.0	0.0	1.3	R5.1	3.4	13.6	R85.0	89.4
Other	70.1	R228.3	233.0	0.0	0.4	0.1	61.8	R98.5	101.7	135.3	R338.6	346.1
World	**5,589.0**	**R8,583.9**	**8,813.9**	**684.4**	**R2,315.3**	**2,396.0**	**1,736.8**	**R2,566.9**	**2,614.8**	**R8,037.9**	**R13,673.9**	**14,056.3**

[1] Excludes pumped storage, except for the United States.
[2] Geothermal, wood, other biomass, waste, solar, wind, hydrogen, sulfur, batteries, and chemicals are included in total.
R=Revised. P=Preliminary. — = Not applicable.
Notes: Data include both electric utility and nonutility sources. Totals may not equal sum of components due to independent rounding.

SOURCE: "Table 11.15. World Net Generation of Electricity by Type, 1980, 1998, and 1999," in *Annual Energy Review 2000*, U.S. Department of Energy, Energy Information Administration, Washington, DC, 2001

TABLE 8.4

World net energy consumption, 1991–2000

Region[1] Country	1991	1992	1993	1994	1995	1996	1997	1998	1999	2000
North America										
Canada	444.4	450.3	454.7	464.2	474.8	486.6	487.7	484.8	497.4	499.8
Mexico	110.5	114.1	119.2	128.9	133.7	143.7	156.0	161.5	171.2	182.8
United States	2,873.0	2,885.1	2,988.4	3,075.5	3,162.4	3,250.1	3,294.6	3,424.0	3,500.9	3,621.0
Other	0.7	0.7	0.7	0.7	0.7	0.7	0.7	0.7	0.8	0.8
Total	**3,428.7**	**3,450.2**	**3,562.9**	**3,669.4**	**3,771.6**	**3,881.1**	**3,939.1**	**4,071.1**	**4,170.4**	**4,304.4**
Central & South America										
Argentina	49.6	56.7	60.5	63.1	68.7	70.6	76.1	82.8	75.1	80.8
Bolivia	2.2	2.4	2.4	2.4	2.9	2.9	2.9	3.2	3.4	3.6
Brazil	242.1	246.3	259.4	271.7	288.2	307.2	322.7	334.3	344.0	360.6
Chile	17.8	20.7	22.1	23.3	24.1	26.0	29.5	30.4	35.1	37.9
Colombia	34.0	31.1	35.4	38.5	41.8	39.9	41.3	42.2	40.4	40.3
Costa Rica	3.6	3.8	4.1	4.7	4.3	4.5	5.1	5.1	5.3	5.9
Cuba	11.3	9.9	9.0	10.0	10.9	11.6	12.4	12.4	13.4	13.8
Dominican Republic	3.4	5.0	5.2	5.5	5.8	6.0	6.5	6.7	6.8	8.8
Ecuador	6.3	6.5	6.8	7.5	7.7	8.4	8.7	9.0	9.4	9.7
El Salvador	1.8	1.9	2.3	2.7	2.8	2.4	3.1	3.1	3.6	4.1
Guatemala	2.5	2.5	2.6	2.7	3.0	3.2	3.0	4.0	4.2	4.8
Honduras	2.1	2.2	2.3	2.5	2.5	2.9	2.3	3.1	2.8	3.6
Jamaica	1.9	2.0	3.3	4.2	5.1	5.3	5.5	5.7	5.8	6.3
Nicaragua	1.4	1.4	1.4	1.5	1.6	1.7	1.9	2.3	2.0	2.2
Panama	3.3	3.5	3.9	4.2	3.9	3.8	4.1	4.4	4.5	4.7
Peru	12.9	11.9	13.5	13.5	16.1	15.7	16.3	16.9	17.4	18.3
Puerto Rico	13.8	14.4	15.0	15.7	16.2	16.6	17.4	17.9	18.7	19.1
Suriname	1.6	1.6	1.6	1.6	1.7	1.7	1.9	1.9	1.8	1.3
Trinidad and Tobago	3.3	3.5	3.3	3.6	3.8	4.0	4.4	4.5	4.6	4.8
Uruguay	4.6	4.8	5.1	5.4	5.8	6.0	6.4	6.6	5.8	7.4
Venezuela	57.2	60.8	62.5	64.3	66.5	68.4	70.8	68.9	68.8	75.1
Virgin Islands, U.S	0.9	0.9	0.9	0.9	0.9	0.9	0.9	0.9	0.9	1.0
Other	5.9	6.2	6.9	7.0	7.7	7.7	8.5	9.5	10.2	10.4
Total	**483.4**	**499.7**	**529.7**	**556.4**	**591.9**	**617.6**	**651.6**	**675.7**	**683.6**	**724.3**
Western Europe										
Austria	47.9	47.8	47.0	47.4	49.2	50.7	51.0	52.1	53.2	54.8
Belgium	61.5	63.8	64.7	67.7	69.8	71.3	72.7	74.6	75.3	78.1
Denmark	29.9	30.9	31.2	30.9	32.2	32.5	32.7	32.9	33.2	33.9
Finland	58.8	64.1	67.1	70.4	69.5	70.5	75.4	80.2	80.3	82.0
France	348.3	354.8	356.0	359.2	366.6	383.9	378.5	394.4	401.4	408.5
Germany	474.8	468.4	465.2	469.6	479.3	485.4	485.6	492.3	493.4	501.7
Greece	32.0	33.4	34.3	35.9	37.2	38.7	40.4	42.3	43.7	46.1
Iceland	4.1	4.2	4.3	4.4	4.6	4.7	5.1	5.8	6.6	7.0
Ireland	13.1	13.8	14.2	14.8	15.4	16.4	17.3	18.5	19.5	20.8
Italy	228.2	232.1	233.7	239.9	247.1	249.2	257.0	265.5	272.4	283.7
Luxembourg	4.8	4.6	4.7	5.1	5.5	5.3	5.6	5.8	5.9	6.2
Netherlands	75.0	77.5	78.9	81.6	83.6	86.8	91.6	94.8	97.7	100.7
Norway	99.2	99.3	102.6	103.4	106.0	105.4	106.0	110.8	110.5	112.5
Portugal	27.2	28.3	28.5	29.7	31.2	32.7	34.3	35.8	38.1	41.1
Spain	137.1	139.6	139.4	145.1	152.0	156.4	166.5	176.7	190.3	201.2
Sweden	132.4	131.3	132.4	130.5	133.9	133.2	134.5	134.5	132.4	139.2
Switzerland	48.6	48.4	48.0	48.5	50.1	50.3	49.8	50.8	52.3	52.6
Turkey	54.0	60.0	65.8	69.4	76.6	84.9	94.6	102.2	105.6	114.2
United Kingdom	294.5	295.0	299.3	301.1	309.1	321.1	319.7	335.1	340.3	345.0
Former Yugoslavia	65.5	—	—	—	—	—	—	—	—	—
Bosnia and Herzegovina	—	10.9	10.9	1.9	2.3	2.4	2.4	2.5	2.6	2.6
Croatia	—	10.9	10.7	11.1	11.5	12.1	12.7	14.1	13.4	12.6
Macedonia, TFYR	—	5.6	5.2	5.3	5.3	5.7	5.8	6.0	5.8	5.9
Slovenia	—	8.9	8.8	11.1	9.4	9.5	9.8	10.4	10.3	10.6
Yugoslavia	—	33.6	29.7	30.7	33.0	33.7	35.8	33.6	31.6	31.5
Other	1.5	1.6	1.5	1.6	1.7	1.7	1.7	1.7	1.8	1.9
Total	**2,238.4**	**2,268.6**	**2,284.1**	**2,316.2**	**2,381.9**	**2,444.4**	**2,486.6**	**2,573.5**	**2,617.7**	**2,694.5**

TABLE 8.4

World net energy consumption, 1991–2000 [CONTINUED]

Region[1] Country	1991	1992	1993	1994	1995	1996	1997	1998	1999	2000
Eastern Europe & Former U.S.S.R.										
Albania	2.7	2.6	3.1	3.6	4.1	5.6	5.0	5.1	5.3	5.4
Bulgaria	36.3	33.9	33.6	33.5	36.7	37.7	33.8	33.1	31.8	34.4
Former Czechoslovakia	76.4	70.9	—	—	—	—	—	—	—	—
Czech Republic	—	—	49.6	51.2	54.0	56.1	55.8	54.8	53.7	54.7
Slovakia	—	—	23.9	24.3	24.8	26.5	26.0	25.7	24.6	25.2
Hungary	33.7	31.2	31.4	31.5	32.3	33.0	33.2	33.5	33.8	35.1
Poland	115.7	112.3	115.1	116.2	119.3	122.6	123.4	120.2	118.0	119.3
Romania	55.2	52.1	51.0	49.6	52.9	54.8	51.0	48.0	44.4	45.7
Former U.S.S.R	1,475.3	—	—	—	—	—	—	—	—	—
Armenia	—	8.3	5.7	5.1	4.7	5.4	5.3	5.4	5.6	4.9
Azerbaijan	—	16.7	16.9	15.7	15.4	15.2	15.8	16.1	17.0	16.7
Belarus	—	39.3	35.2	31.4	29.4	28.9	30.6	28.1	30.3	26.8
Estonia	—	7.1	6.6	6.9	6.8	7.1	7.1	7.1	6.6	5.4
Georgia	—	11.3	9.9	7.1	7.0	6.7	6.7	7.3	7.4	7.9
Kazakhstan	—	86.2	83.4	64.9	64.3	57.2	49.5	47.2	44.8	48.3
Kyrgyzstan	—	8.8	9.3	9.4	12.6	10.6	9.9	10.0	10.0	9.8
Latvia	—	7.5	6.1	5.9	5.9	6.0	5.9	5.8	5.5	5.2
Lithuania	—	11.2	10.3	9.8	9.7	8.2	8.8	9.2	9.4	6.9
Moldova	—	9.8	6.9	7.8	7.2	6.9	6.6	6.2	5.0	3.7
Russia	—	881.5	832.2	732.9	740.5	730.5	720.6	714.6	730.8	767.1
Tajikistan	—	16.3	15.1	14.8	14.3	14.1	13.9	13.9	14.2	12.5
Turkmenistan	—	8.6	7.9	6.6	6.5	6.0	6.6	5.2	6.3	7.7
Ukraine	—	216.7	200.6	178.1	168.8	159.5	156.7	151.2	147.1	151.7
Uzbekistan	—	44.2	40.4	43.7	40.5	44.0	41.5	41.6	41.9	41.9
Total	**1,795.3**	**1,676.4**	**1,594.1**	**1,450.0**	**1,457.5**	**1,442.6**	**1,413.6**	**1,389.4**	**1,393.5**	**1,436.2**
Middle East										
Bahrain	3.1	3.4	3.7	4.0	4.0	4.4	4.4	5.0	5.2	5.4
Cyprus	1.8	2.1	2.3	2.3	2.2	2.3	2.4	2.6	2.7	2.9
Iran	56.4	60.4	66.9	72.0	74.6	79.8	85.8	90.8	98.7	111.9
Iraq	17.7	22.1	23.0	24.5	25.4	25.5	25.9	26.6	26.0	25.4
Israel	18.4	21.2	22.4	24.4	25.7	27.4	29.6	32.1	32.9	34.9
Jordan	3.4	4.0	4.3	4.6	5.5	5.6	5.9	6.3	6.8	7.1
Kuwait	9.4	14.7	17.6	19.9	20.7	22.3	23.4	26.2	27.6	29.0
Lebanon	2.7	3.3	4.2	4.7	5.0	6.8	8.1	8.0	8.4	8.6
Oman	4.8	4.5	5.1	5.4	5.6	5.9	6.4	7.2	7.4	7.5
Qatar	4.1	4.5	4.8	5.1	5.2	5.7	6.0	7.1	7.8	8.6
Saudi Arabia	64.4	68.8	76.4	84.6	91.0	94.0	100.0	106.6	110.7	114.9
Syria	11.0	11.4	11.4	13.7	13.7	15.1	16.0	17.4	18.2	17.7
United Arab Emirates	15.1	15.3	15.4	16.5	21.8	23.2	24.9	29.2	32.5	36.0
Yemen	1.6	1.7	1.8	1.9	2.1	2.0	2.2	2.2	2.6	3.0
Total	**213.9**	**237.5**	**259.5**	**283.7**	**302.5**	**320.2**	**341.0**	**367.2**	**387.5**	**412.8**
Africa										
Algeria	14.5	15.1	15.8	16.3	16.5	17.6	18.6	20.1	20.9	21.8
Angola	1.7	1.7	1.7	1.7	1.7	1.7	1.7	1.8	1.4	1.1
Cameroon	2.5	2.5	2.5	2.5	2.6	2.7	2.9	2.9	3.2	3.4
Congo (Kinshasa)	4.7	5.4	4.1	4.1	5.5	5.9	5.7	4.8	4.4	4.6
Cote d'Ivoire (Ivory Coast)	1.7	1.5	2.0	2.0	2.7	2.9	3.6	3.0	3.1	3.6
Egypt	39.6	40.5	44.4	46.6	48.4	48.1	51.0	55.1	60.2	64.7
Ghana	5.4	5.4	5.4	5.3	5.3	5.7	6.0	4.7	4.7	5.5
Kenya	3.1	3.1	3.4	3.5	3.6	3.8	4.0	4.1	4.2	4.4
Libya	.8	14.8	14.9	15.6	15.7	16.0	16.6	17.0	17.5	18.0
Morocco	8.7	9.5	10.0	10.8	11.5	11.7	12.4	13.2	13.8	14.3
Nigeria	12.6	13.1	12.8	13.7	12.9	13.4	13.7	14.0	14.4	14.8
South Africa	146.1	144.6	149.4	156.2	160.9	168.3	175.6	175.8	176.0	181.5
Tunisia	5.0	5.4	5.5	5.9	6.5	7.1	7.6	7.9	8.7	9.6
Zambia	5.7	5.7	5.7	5.7	5.7	5.9	6.2	6.1	5.6	5.8
Zimbabwe	9.7	8.6	8.8	7.6	7.9	8.8	9.9	7.4	10.6	10.5
Other	17.2	17.7	19.4	19.7	20.3	21.7	22.4	21.9	22.3	23.0
Total	**293.0**	**294.6**	**305.6**	**317.1**	**327.8**	**341.5**	**357.8**	**359.9**	**371.0**	**386.6**

TABLE 8.4

World net energy consumption, 1991–2000 [CONTINUED]

Region[1] Country	1991	1992	1993	1994	1995	1996	1997	1998	1999	2000
Asia & Oceania										
Afghanistan	1.0	0.8	0.7	0.8	0.7	0.6	0.6	0.5	0.5	0.5
American Samoa	0.1	0.1	0.1	0.1	0.1	0.1	0.1	0.1	0.1	0.1
Australia	138.2	142.3	146.1	149.5	154.5	158.4	163.3	175.3	181.5	188.5
Bangladesh	7.8	8.4	8.7	9.3	10.5	10.9	11.2	10.8	11.3	12.5
Bhutan	0.1	0.1	0.1	0.1	(s)	0.3	0.4	0.2	0.4	0.4
Brunei	1.2	1.2	1.4	1.5	1.7	1.9	2.1	2.2	2.1	2.1
Burma	2.4	2.7	3.0	3.2	3.6	3.5	4.0	3.6	4.2	4.4
Cambodia	0.1	0.2	0.2	0.2	0.2	0.2	0.2	0.1	0.1	0.1
China	600.9	670.6	744.1	816.5	883.4	927.2	987.5	1,014.8	1,083.7	1,206.3
Fiji	0.4	0.4	0.4	0.5	0.5	0.5	0.5	0.5	0.5	0.5
French Polynesia	0.3	0.3	0.3	0.3	0.3	0.3	0.4	0.4	0.4	0.4
Guam	0.7	0.7	0.7	0.7	0.7	0.7	0.7	0.7	0.8	0.8
Hong Kong	24.7	25.6	27.3	25.4	28.8	31.9	32.6	33.9	34.1	35.4
India	280.6	295.1	316.9	341.9	369.7	385.4	411.7	395.9	423.4	509.9
Indonesia	45.4	46.0	41.2	46.0	52.6	55.1	66.0	67.6	76.1	86.1
Japan	791.3	797.9	806.4	857.9	881.4	899.0	927.3	926.7	945.4	943.7
Korea, North	48.2	34.3	34.3	33.4	32.5	31.6	30.4	28.9	28.6	31.1
Korea, South	103.0	113.6	125.4	159.4	176.3	196.3	214.2	209.8	232.8	254.1
Laos	0.2	0.2	0.3	0.5	0.3	0.4	0.4	0.3	0.5	0.7
Malaysia	24.9	25.8	30.6	34.5	40.1	45.1	50.8	53.2	57.2	58.6
Macau	0.9	1.0	1.1	1.2	1.5	1.4	1.4	1.4	1.4	1.5
Mongolia	2.9	2.7	2.3	2.5	2.7	2.7	2.7	2.6	2.6	2.7
Nepal	0.8	0.9	0.9	1.0	1.2	1.2	1.1	1.3	1.3	1.4
New Caledonia	1.0	1.0	1.0	1.1	1.5	1.4	1.5	1.5	1.4	1.5
New Zealand	30.8	29.4	31.2	32.6	32.9	34.0	33.9	34.8	35.0	33.3
Pakistan	36.8	40.7	43.5	45.3	47.9	50.8	52.7	55.4	58.1	58.3
Papua New Guinea	1.6	1.6	1.6	1.6	1.6	1.6	1.6	1.7	1.7	1.5
Philippines	22.0	21.2	21.9	27.0	29.6	32.5	35.2	36.7	34.9	37.8
Samoa	(s)	0.1	0.1	0.1	0.1	0.1	0.1	0.1	0.1	0.1
Singapore	14.8	15.5	16.5	18.1	19.4	20.5	22.9	24.2	25.1	25.9
Sri Lanka	3.1	2.9	3.7	4.0	4.4	4.1	4.7	5.2	5.6	6.2
Taiwan	84.6	90.2	98.3	106.2	105.3	111.6	118.0	127.4	129.9	139.3
Thailand	44.7	50.7	56.1	63.0	70.9	77.4	81.5	77.6	81.0	90.3
U.S. Pacific Islands	0.2	0.2	0.2	0.2	0.2	0.2	0.2	0.2	0.2	0.2
Vietnam	8.4	8.8	9.7	11.2	13.3	15.4	17.3	19.5	21.2	24.0
Other	0.1	0.1	0.2	0.2	0.2	0.2	0.2	0.2	0.2	0.3
Total	2,324.5	2,433.2	2,576.3	2,796.6	2,970.6	3,104.4	3,279.2	3,315.4	3,483.5	3,760.2
World Total	10,777.2	10,860.3	11,112.2	11,389.4	11,803.9	12,151.7	12,468.8	12,752.1	13,107.3	13,719.1

[1] Preliminary.

--= Not applicable.

(s) = Value less than 50 million kilowatthours.

Notes: Sum of components may not equal total due to independent rounding.

Consumption equals generation plus imports minus exports minus distribution losses.

SOURCE: "Table 6.2. World Total Net Electricity Consumption, 1991–2000," in *International Energy Annual 2000,* U.S. Department of Energy, Energy Information Administration, Washington, DC, 2002

FIGURE 8.8

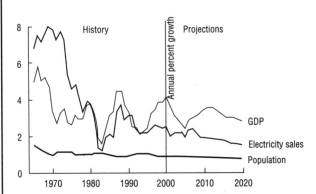

Population, gross domestic product, and electricity sales, 1965–2020

(5-year moving average annual percent growth)

SOURCE: "Figure 45. Population, gross domestic product, and electricity sales, 1965–2020 (5–year moving average annual percent growth)," in *Annual Energy Outlook 2002,* U.S. Department of Energy, Energy Information Administration, Washington, DC, 2001

CHAPTER 9
ENERGY CONSERVATION

ENERGY CONSERVATION AND EFFICIENCY

Energy conservation is the efficient use of energy, without necessarily curtailing the services that energy provides. Conservation occurs when societies develop efficient technologies that reduce energy needs. Environmental concerns, such as acid rain and the potential for global warming, have increased public awareness about the importance of energy conservation.

Energy efficiency can be measured by two indicators. The first is energy consumption per person (per capita) per year. Annual per person energy consumption in the United States was 215 million Btu in 1949, topped out at 361 million Btu in 1978 and 1979, dropped to 314 million Btu in 1983, and rose again to 350 million Btu in 2000. (See Figure 9.1.)

The second indicator of efficiency is energy consumption per dollar of gross domestic product (GDP). GDP is the total value of goods and services produced by a nation. When a country grows in its energy efficiency, it uses less energy to produce the same amount of goods and services. In 1950 nearly 21,000 Btu of energy were consumed for each dollar of GDP. (See Figure 9.2.) In 1970 about 19,000 Btu of energy were consumed per dollar, and 10,570 Btu were used per dollar in 2000.

ENERGY CONSERVATION, PUBLIC HEALTH, AND THE ENVIRONMENT

In the coming decades, global environmental issues could significantly affect patterns of energy use around the world.

—*International Energy Outlook 2000,* Energy Information Administration, Washington, D.C., March 2000

The connection between energy policy and the health of both people and the environment has become clearer in recent years. People living in cities with high levels of pollution have higher risks of mortality and certain diseases

FIGURE 9.1

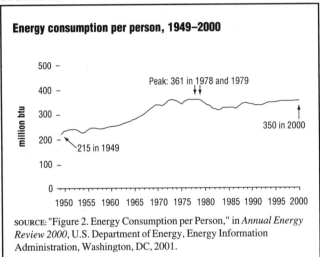

Energy consumption per person, 1949–2000

SOURCE: "Figure 2. Energy Consumption per Person," in *Annual Energy Review 2000*, U.S. Department of Energy, Energy Information Administration, Washington, DC, 2001.

FIGURE 9.2

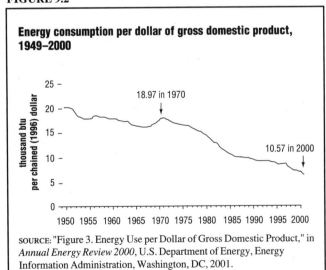

Energy consumption per dollar of gross domestic product, 1949–2000

SOURCE: "Figure 3. Energy Use per Dollar of Gross Domestic Product," in *Annual Energy Review 2000*, U.S. Department of Energy, Energy Information Administration, Washington, DC, 2001.

TABLE 9.1

Air pollutants, health risks, and contributing sources

Pollutants	Health risks	Contributing sources
Ozone[1] (O_3)	Asthma, reduced respiratory function, eye irritation	Cars, refineries, dry cleaners
Particulate matter (PM-IO)	Bronchitis, cancer, lung damage	Dust, pesticides
Carbon monoxide (CO)	Blood oxygen carrying capacity reduction, cardiovascular and nervous system impairments	Cars, power plants, wood stoves
Sulphur dioxide (SO_2)	Respiratory tract impairment, destruction of lung tissue	Power plants, paper mills
Lead (Pb)	Retardation and brain damage, esp. children	Cars, nonferrous smelters, battery plants
Nitrogen dioxide (NO_2)	Lung damage and respiratory illness	Power plants, cars, trucks

[1] Ozone refers to tropospheric ozone which is hazardous to human health.

SOURCE: Fred Seitz and Christine Plepys, "Table 1. Criteria air pollutants, health risks and sources," in *Healthy People 2000: Statistical Notes*, Centers for Disease Control and Prevention, Atlanta, GA, 1995

than those living in less polluted cities. Energy-related emissions generate a vast majority of these polluting chemicals. (Table 9.1 shows some air pollutants and their sources.) In the past few years a number of major studies have documented a link between air pollution and public health. Air pollution has been related to such diseases as asthma, bronchitis, emphysema, and lung cancer. The American Lung Association estimates the annual health costs of exposure to the most serious air pollutants to be in the billions. Clean and efficient energy technologies represent a cost-effective investment in public health.

Evidence of global warming and its consequences has also accumulated rapidly in recent years. Global warming is long-term climate change—a worldwide temperature increase—caused by the greenhouse effect. The greenhouse effect is the blocking of outward heat radiation from the atmosphere by high levels of gases called "greenhouse gases": carbon dioxide, methane, nitrogen oxide, hydrofluorocarbons, sulfur dioxides, and perfluorocarbons. Just as the glass of a greenhouse or the windows of a car trap heat, greenhouse gases keep the earth warmer than if the atmosphere contained only oxygen and nitrogen.

The United Nations' Intergovernmental Panel on Climate Change (IPCC) Examines Global Warming

The United Nations' Intergovernmental Panel on Climate Change (IPCC), a group of 2,000 of the world's leading scientists, was established in 1988 and reported in 1995 that global warming is real, serious, and accelerating. It determined that the most likely cause is primarily

humans burning coal, oil, and gasoline, which has increased the amount of carbon dioxide and other greenhouse gases in the Earth's atmosphere. Deforestation also contributes by reducing the Earth's ability to absorb carbon dioxide from burning fossil fuels.

The IPCC published its *Third Assessment Report* in 2001. This report notes that scientists now have a clearer understanding of the causes and consequences of global warming because of the intensive climate research and environmental monitoring that has been conducted. They describe the effect that global warming will have on weather patterns, water resources, the seasons, ecosystems, and extreme climate events, and they urge governments to move forward quickly with climate change policies.

The Kyoto Protocol

In December 1997 the United Nations convened a 160-nation conference on global warming in Kyoto, Japan, in hopes of producing a new treaty on climate change that would place binding caps on industrial emissions. The treaty, called the Kyoto Protocol to the United Nations Framework Convention on Climate Change (or simply the Kyoto Protocol), binds industrialized nations to reducing their emissions of the six greenhouse gases below 1990 levels by 2012, with each country having a different target.

Levels of greenhouse gas emissions for the United States from 1990 to 1998 are shown in Table 9.2 in million metric tons of carbon equivalents (MMTCE). Under the terms of initial drafts of the protocol, the United States was to cut emissions by 7 percent, most European nations by 8 percent, and Japan by 6 percent. Reductions were to begin by 2008 and to be achieved by 2012. Developing nations were not required to make such pledges.

The United States had proposed a program of voluntary pledges by developing nations, but that section was deleted, as was a tough system of enforcement. Instead, each country was to decide for itself how to achieve its goal. The treaty would provide market-driven tools, such as buying and selling credits, for reducing emissions. It would also set up a Clean Development Fund to help poorer nations with technology to reduce their emissions.

The Kyoto Protocol would mark the first time nations have made such sweeping pledges to cut emissions, but there has been difficulty getting ratification. In the United States, President Bill Clinton signed the protocol but the Senate did not ratify it. By early 2001 the Kyoto Protocol was near collapse. However, diplomats from 178 nations met in Bonn, Germany, in July 2001 and drafted a compromise to preserve the global-warming treaty. In October 2001, 2,000 delegates from 160 countries began 12 days of talks in Marrakesh, Morocco, to complete a final draft of the Kyoto Protocol. President George W. Bush, though, stood firm and rejected the Kyoto Protocol, characterizing

TABLE 9.2

Trends in greenhouse gas emissions and sinks, 1990–98

Gas/Source	1990	1991	1992	1993	1994	1995	1996	1997	1998
CO_2	**1,340.3**	**1,326.1**	**1,350.4**	**1,383.3**	**1,404.8**	**1,416.5**	**1,466.2**	**1,486.4**	**1,494.0**
Fossel Fuel Combustion	1,320.1	1,305.8	1,330.1	1,361.5	1,382.0	1,392.0	1,441.3	1,460.7	1,468.2
Cement Manufacture	9.1	8.9	8.9	9.4	9.8	10.0	10.1	10.5	10.7
Naural Gas Flaring	2.5	2.8	2.8	3.7	3.8	4.7	4.5	4.2	3.9
Lime Manufacture	3.0	3.0	3.1	3.1	3.2	3.4	3.6	3.7	3.7
Waste Combustion	2.8	3.0	3.0	3.1	3.1	3.0	3.1	3.4	3.5
Limestone Dolomite Use	1.4	1.3	1.2	1.1	1.5	1.9	2.0	2.3	2.4
Soda Ash Manufacture and Consumption	1.1	1.1	1.1	1.1	1.1	1.2	1.2	1.2	1.2
Carbon Dioxide Consumption	0.2	0.2	0.2	0.2	0.2	0.3	0.3	0.4	0.4
Land-Use Change and Forestry (Sink)[a]	(316.4)	(316.3)	(316.2)	(212.7)	(211.3)	(211.8)	(211.3)	(211.1)	(210.8)
International Bunker Fuels[b]	32.2	32.7	30.0	27.2	26.7	27.5	27.9	29.9	31.3
CH_4	**177.9**	**177.7**	**179.4**	**178.7**	**181.6**	**184.1**	**183.1**	**183.8**	**180.9**
Landfills	58.2	58.1	59.1	59.6	59.9	60.5	60.2	60.2	58.8
Enteric Fermentation	32.7	32.8	33.2	33.7	34.5	34.9	34.5	34.2	33.7
Natural Gas Systems	33.0	33.4	33.9	34.6	34.3	34.0	34.6	34.1	33.6
Manure Management	15.0	15.5	16.0	17.1	18.8	19.7	20.4	22.1	22.9
Coal Mining	24.0	22.8	22.0	19.2	19.4	20.3	18.9	18.8	17.8
Petroleum Systems	7.4	7.5	7.2	6.9	6.7	6.7	6.5	6.5	6.3
Rice Cultivation	2.4	2.3	2.6	2.4	2.7	2.6	2.4	2.6	2.7
Stationary Sources	2.3	2.4	2.4	2.4	2.4	2.5	2.6	2.3	2.3
Mobil Sources	1.5	1.5	1.5	1.5	1.5	1.4	1.4	1.4	1.3
Wastewater Treatment	0.9	0.9	0.9	0.9	0.9	0.9	0.9	0.9	0.9
Petrochemical Production	0.3	0.3	0.3	0.4	0.4	0.4	0.4	0.4	0.4
Agricultural Residue Burning	0.2	0.2	0.2	0.1	0.2	0.2	0.2	0.2	0.2
Silicon Carbide Production	+	+	+	+	+	+	+	+	+
International Bunker Fuels[b]	+	+	+	+	+	+	+	+	+
N_2O	**108.2**	**110.5**	**113.3**	**113.8**	**121.5**	**118.8**	**121.5**	**122.4**	**119.4**
Agricultural Soil Management	75.3	76.3	78.2	77.3	83.5	80.4	82.4	84.2	83.9
Mobile Sources	13.8	14.6	15.7	16.5	17.1	17.4	17.5	17.3	17.2
Nitric Acid	4.9	4.9	5.0	5.1	5.3	5.4	5.6	5.8	5.8
Stationary Sources	3.8	3.8	3.9	3.9	4.0	4.0	4.2	4.2	4.3
Manure Management	3.4	3.6	3.5	3.7	3.8	3.7	3.8	3.9	4.0
Human Sewage	2.0	2.0	2.0	2.0	2.1	2.1	2.1	2.1	2.2
Adipic Acid	5.0	5.2	4.8	5.2	5.5	5.5	5.7	4.7	2.0
Agricultural Residue Burning	0.1	0.1	0.1	0.1	0.1	0.1	0.1	0.1	0.1
Waste Combustion	0.1	0.1	0.1	0.1	0.1	0.1	0.1	0.1	0.1
International Bunker Fuels[b]	0.3	0.3	0.3	0.2	0.2	0.2	0.2	0.3	0.3
HFCs, PFCs, and SF_6	**23.3**	**22.0**	**23.5**	**23.8**	**25.1**	**29.0**	**33.5**	**35.3**	**40.3**
Substitution of Ozone Depleting Substances	0.3	0.2	0.4	1.4	2.7	7.0	9.9	12.3	14.5
HCFC-22 Production	9.5	8.4	9.5	8.7	8.6	7.4	8.5	8.2	10.9
Electrical Transmission and Distribution	5.6	5.9	6.2	6.4	6.7	7.0	7.0	7.0	7.0
Magnesium Production and Processing	1.7	2.0	2.2	2.5	2.7	3.0	3.0	3.0	3.0
Aluminum Production	5.4	4.7	4.4	3.8	3.2	3.1	3.2	3.0	2.8
Semiconductor Manufacture	0.8	0.8	0.8	1.0	1.1	1.5	1.9	1.9	2.1
Total Emissions	**1,649.7**	**1,636.2**	**1,666.6**	**1,699.7**	**1,733.0**	**1,748.5**	**1,804.4**	**1,827.9**	**1,834.6**
Net Emission (Sources and Sinks)	**1,333.3**	**1,320.0**	**1,350.5**	**1,487.0**	**1,520.7**	**1,536.6**	**1,593.1**	**1,616.8**	**1,623.8**

+Does not exceed 0.05 MMTCE (Million metric tons carbon equivalent)
[a]Sinks are only included in net emissions total. Estimates of net carbon sequestration due to land-use change and forestry activities exclude non-forest soils, and are based partially upon projections of forest carbon stocks.
[b]Emissions from International Bunker Fuels are not included in totals.
Note: Totals may not sum due to independent rounding.

SOURCE: "Table ES-1: Recent Trends in U.S. Greenhouse Gas Emissions and Sinks (MMTCE)," in *Inventory of U.S. Greenhouse Gas Emissions and Sinks: 1990–1998*, U.S. Environmental Protection Agency, Washington, DC, April 2000

it as "fatally flawed" and saying implementing it would harm the U.S. economy and unfairly require only the industrial nations to cut emissions. As of November 2002 the United States had not signed the protocol.

U.S. feelings on the Kyoto Protocol are widely varied. Business leaders believe the treaty would go too far, while environmentalists believe the Kyoto standards would not go far enough. Some experts doubt that any action emerging from Kyoto would be sufficient to prevent the doubling of greenhouse gases. Representatives of the oil industry and business community contend the treaty would spell economic pain for the United States. The fossil-fuel industry

TABLE 9.3

Transportation conservation options

Improve the technical efficiency of vehicles
1. Higher fuel economy requirements—CAFE standards (R)
2. Reducing congestion: smart highways (E,I), flextime (E,R), better signaling (I), improved maintenance of roadways (I), time of day charges (E), improved air traffic controls (I,R), plus options that reduce vehicular traffic
3. Higher fuel taxes (E)
4. Gas guzzler taxes, or feebate schemes (E)
5. Support for increased R&D (EJ)
6. Inspection and maintenance programs (R)

Increase load factor
1. HOV lanes (I)
2. Forgiven tolls (E), free parking for carpools (E)
3. Higher fuel taxes (E)
4. Higher charges on other vmt trip-dependent factors (E): parking (taxes, restrictions, end of tax treatment as business cost), tolls, etc.

Change to more efficient modes
1. Improvements in transit service
   a. New technologies—maglev, high speed trains (EJ)
   b. Rehabilitation of older systems (I)
   c. Expansion of service—more routes, higher frequency (I)
   d. Other service improvements (I)—dedicated busways, better security, more bus stop shelters, more comfortable vehicles
2. Higher fuel taxes (E)
3. Reduced transit fares through higher US. transit subsidies (E)ᵃ
4. Higher charges on other vmt/trip-dependent factors for less efficient modes (E)—tolls, parking
5. Shifting urban form to higher density, more mixed use, greater concentration through zoning changes (R), encouragement of "infill" development (E,R,I), public investment in infrastructure (I), etc.

Reduce number or length of trips
1. Shifting urban form to higher density, more mixed use, greater concentration (E,R,I)
2. Promoting working at home or at decentralized facilities (EJ)
3. Higher fuel taxes (E)
4. Higher charges on other vmt/trip-dependent factors (E)

Shift to alternative fuels
1. Fleet requirements for alternative fuel-capable vehicles and actual use of alternative fuels (R)
2. Low-emission/zero emission vehicle (LEV/ZEV) requirements (R)
3. Various promotions (E): CAFE credits, emission credits, tax credits, etc.
4. Higher fuel taxes that do not apply to alternative fuels (E), or subsidies for the alternatives (E)
5. Support for increased R&D (EJ)
6. Public investment—government fleet investments (I)

Freight options
1. RD&D of technology improvements (E,I)

ᵃU.S. transit subsidies, already among the highest in the developed world, may merely promote inefficiencies.

KEY: CAFE = corporate average fuel economy; E = economic incentive; HOV = high-occupancy vehicle; I = public investment; maglev = trains supported by magnetic levitation; R = regulatory action; RD&D = research, development, and demonstration; vmt = vehicle-miles traveled.

SOURCE: "Table 5-1: Transportation conservation options," in *Saving Energy in U.S. Transportation*, Office of Technology Assessment, Washington, DC, 1994

and conservative politicians portray the protocol as unworkable and too costly to the American economy.

Despite the lack of support from the United States, the Kyoto Protocol still has a chance to go into effect worldwide. To take effect the treaty must be ratified by 55 of the nations responsible for 55 percent of industrialized nations' carbon dioxide emissions in 1990. By November 2002 more than 95 countries, including all the European Union members, Japan, and Norway, had ratified it. Russia, New Zealand, and Canada were expected to ratify the treaty as well, which would bring the treaty into effect.

Although President Bush rejected the Kyoto Protocol, he promised to address the issue of greenhouse gas emissions and global warming. He proposed studying the problem and funding new technologies to reduce carbon dioxide emissions.

EFFICIENCY IN THE TRANSPORTATION SECTOR

The U.S. transportation system plays a central role in the economy. Highway transportation is dependent on internal-combustion-engine vehicles fueled almost exclusively by petroleum. The Energy Information Administration (EIA) of the U.S. Department of Energy (DOE), in its *Annual Energy Review 2000* (2001), notes that the transportation sector accounted for 27 percent of all energy consumed in the United States in 2000. Figure 1.10 in Chapter 1 shows that in 2000 Americans used 26.6 quadrillion Btu of energy for transportation, of which petroleum made up 97 percent. Despite improvements in transportation efficiency in recent decades, the United States still consumes more than one-third of the world's transport energy. In 1997 the United States imported more oil than it produced for the first time in history and continued to do so in later the years. American dependence on oil not only makes the economy vulnerable to supply and price volatility in the world oil market but also leads to air pollution problems.

Automotive Efficiency

Policymakers interested in transportation energy conservation have an array of conservation options. (See Table 9.3.) However, not all options are mutually supportive. For example, efforts to promote a freer flow of automobile traffic, such as high-occupancy vehicle (HOV) lanes or free parking for carpools, may sabotage efforts to shift travelers to mass transit or to reduce trip lengths and frequency. Policymakers must consider how the implementation of one strategy will fit into an overall transportation plan. In the United States the automobile dominates the transportation sector, as cars and light-duty vehicles used 63 percent of all transportation energy in 1998. (See Figure 1.13 in Chapter 1.)

Motor gasoline, which is divided among passenger cars, light and heavy-duty trucks, aircraft, and miscellaneous other modes of transportation, consumed approximately two-thirds of the oil used in the United States in 1999, according to the Bureau of Transportation Statistics' *Transportation Statistics Annual Report 2000* (2001). The major growth in fuel use over the past 10 years has been that consumed by trucks, mainly increased sales of sport utility vehicles (SUVs). In 1999 truck fuel use was nearly equal to automobile fuel use. Meanwhile, automobile fuel use has remained fairly constant because of fuel efficiency increases that have offset the growth in car miles traveled. Boosting truck efficiency will become increasingly important in holding down oil demand.

FIGURE 9.3

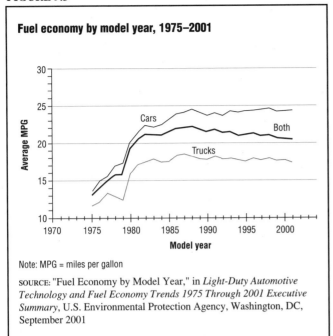

Fuel economy by model year, 1975–2001

Note: MPG = miles per gallon

SOURCE: "Fuel Economy by Model Year," in *Light-Duty Automotive Technology and Fuel Economy Trends 1975 Through 2001 Executive Summary*, U.S. Environmental Protection Agency, Washington, DC, September 2001

FIGURE 9.4

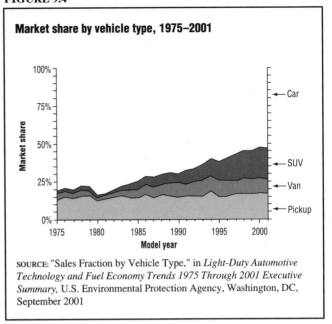

Market share by vehicle type, 1975–2001

SOURCE: "Sales Fraction by Vehicle Type," in *Light-Duty Automotive Technology and Fuel Economy Trends 1975 Through 2001 Executive Summary*, U.S. Environmental Protection Agency, Washington, DC, September 2001

THE CORPORATE AVERAGE FUEL ECONOMY (CAFE) STANDARDS. The 1973 OPEC oil embargo painfully reminded the United States how dependent it had become on foreign sources of fuel. For example, although the United States made up only about 5 percent of the world's population in 2000, it consumed nearly 29 percent of the world's supply of oil, much of which was imported from the Middle East. The oil embargo prompted Congress to pass the 1975 Automobile Fuel Efficiency Act (PL 96-426), which set the initial Corporate Average Fuel Economy (CAFE) standards.

The CAFE standards required domestic automakers to increase the average mileage of new cars sold to 27.5 miles per gallon (mpg) by 1985. Under CAFE rules, car manufacturers can still sell large, less efficient cars, but to meet the average fuel efficiency rates, they also have to sell smaller, more efficient cars. Automakers that failed to meet each year's CAFE standards were fined. Those that managed to surpass the rates earned credits that they could use in years when they fell below the CAFE requirements. Faced with the CAFE standards, the car companies became more inventive and managed to keep their cars relatively large and roomy while increasing mileage with innovations like electronic fuel injection and four valves per cylinder.

The CAFE regulations have had a significant effect on fuel efficiency. In the decades since the first oil shock in 1973, the fuel economy of passenger cars (all cars currently on the road) increased from 13.4 mpg in 1973 to 21.4 mpg in 1999. (See Table 9.4.) Greater gains have been made in the economy of new cars. In 1974, just after the oil embargo, cars averaged 14.2 mpg; in 2001 the average new-car fuel economy was 24.2 mpg. (See Figure

9.3.) The total automobile fleet fuel economy is expected to increase as more fuel-efficient cars enter the market and older, less fuel-efficient autos drop out of the nation's fleet. However, new-car fuel economy has remained virtually unchanged since 1986. Since 1988 nearly all gains in automobile efficiency have been offset by increased weight and power in new vehicles.

The Persian Gulf War in 1991 was another strong reminder to the United States of its continuing heavy dependence on foreign oil, causing some members of Congress to want to raise the CAFE standards to 45 mpg for cars and 35 mpg for light trucks. Those in favor of raising CAFE standards claimed that this would save about 2.8 million barrels of oil per day. They also noted that if cars become even more fuel-efficient in the future, emissions of carbon dioxide would be significantly reduced. The domestic auto industry opposed the bill to raise CAFE standards, and actions to increase automobile efficiency failed in Congress.

SUVs, vans, and pickup trucks make up the "light truck" automotive group, the fastest growing segment of the auto industry. Light trucks accounted for 47 percent of the U.S. light vehicle market in 2001 and produced most of the profits of the major auto companies. (See Figure 9.4.) These types of vehicles fall under less stringent emissions standards than automobiles because of their truck classification, which allows them more lenient EPA regulations. In 1999 the EPA proposed new regulations tightening emissions standards on cars, minivans, SUVs under 8,500 pounds, and small pickup trucks. The regulations would not have affected the rules for big SUVs and pickups, allowing these bigger vehicles to emit up to as much as five times more nitrogen oxides than cars.

TABLE 9.4

Motor vehicle mileage, fuel consumption, and fuel rates, 1949–1999

Year	Passenger cars			Vans, pickup trucks, and sport utility vehicles[1]			Trucks[2]			All motor vehicles[3]		
	Mileage (miles per vehicle)	Fuel consumption (gallons per vehicle)	Fuel rate (miles per gallon)	Mileage (miles per vehicle)	Fuel consumption (gallons per vehicle)	Fuel rate (miles per gallon)	Mileage (miles per vehicle)	Fuel consumption (gallons per vehicle)	Fuel Rate (miles per gallon)	Mileage (miles per vehicle)	Fuel consumption (gallons per vehicle)	Fuel rate (miles per gallon)
1949	[4]9,388	[4]627	[4]15.0	(5)	(5)	(5)	[6]9,712	[6]1,080	[6]9.0	9,498	726	13.1
1950	[4]9,060	[4]603	[4]15.0	(5)	(5)	(5)	[6]10,316	[6]1,229	[6]8.4	9,321	725	12.8
1951	[4]9,186	[4]614	[4]15.0	(5)	(5)	(5)	[6]10,545	[6]1,242	[6]8.5	9,460	735	12.9
1952	[4]9,360	[4]639	[4]14.7	(5)	(5)	(5)	[6]10,769	[6]1,288	[6]8.4	9,642	762	12.7
1953	[4]9,377	[4]640	[4]14.6	(5)	(5)	(5)	[6]10,963	[6]1,283	[6]8.5	9,684	760	12.7
1954	[4]9,349	[4]641	[4]14.6	(5)	(5)	(5)	[6]10,682	[6]1,281	[6]8.3	9,605	758	12.7
1955	[4]9,447	[4]645	[4]14.6	(5)	(5)	(5)	[6]10,576	[6]1,293	[6]8.2	9,661	761	12.7
1956	[4]9,496	[4]654	[4]14.5	(5)	(5)	(5)	[6]10,511	[6]1,309	[6]8.0	9,688	771	12.6
1957	[4]9,348	[4]658	[4]14.2	(5)	(5)	(5)	[6]10,774	[6]1,304	[6]8.3	9,609	773	12.4
1958	[4]9,500	[4]670	[4]14.2	(5)	(5)	(5)	[6]10,768	[6]1,303	[6]8.3	9,732	782	12.4
1959	[4]9,615	[4]674	[4]14.3	(5)	(5)	(5)	[6]10,702	[6]1,328	[6]8.1	9,817	789	12.4
1960	[4]9,518	[4]668	[4]14.3	(5)	(5)	(5)	[6]10,693	[6]1,333	[6]8.0	9,732	784	12.4
1961	[4]9,521	[4]663	[4]14.4	(5)	(5)	(5)	[6]10,537	[6]1,341	[6]7.9	9,708	781	12.4
1962	[4]9,494	[4]662	[4]14.3	(5)	(5)	(5)	[6]10,554	[6]1,337	[6]7.9	9,687	779	12.4
1963	[4]9,587	[4]655	[4]14.6	(5)	(5)	(5)	[6]10,395	[6]1,380	[6]7.5	9,737	780	12.5
1964	[4]9,665	[4]661	[4]14.6	(5)	(5)	(5)	[6]10,408	[6]1,389	[6]7.5	9,805	787	12.5
1965	[4]9,603	[4]661	[4]14.5	(5)	(5)	(5)	[6]10,851	[6]1,387	[6]7.8	9,826	787	12.5
1966	[4]9,733	[4]688	[4]14.1	8,077	833	9.7	12,537	2,250	5.6	9,675	780	12.4
1967	[4]9,849	[4]699	[4]14.1	7,877	801	9.8	12,789	2,294	5.6	9,751	786	12.4
1968	[4]9,922	[4]714	[4]13.9	8,376	849	9.9	12,402	2,240	5.5	9,864	805	12.2
1969	[4]9,921	[4]727	[4]13.6	8,355	851	9.8	13,484	2,459	5.5	9,885	821	12.0
1970	[4]9,989	[4]737	[4]13.5	8,676	866	10.0	13,565	2,467	5.5	9,976	830	12.0
1971	[4]10,097	[4]743	[4]13.6	9,082	888	10.2	14,117	2,519	5.6	10,133	839	12.1
1972	[4]10,171	[4]754	[4]13.5	9,534	922	10.3	14,780	2,657	5.6	10,279	857	12.0
1973	[4]9,884	[4]737	[4]13.4	9,779	931	10.5	15,370	2,775	5.5	10,099	850	11.9
1974	[4]9,221	[4]677	[4]13.6	9,452	862	11.0	14,995	2,708	5.5	9,493	788	12.0
1975	[4]9,309	[4]665	[4]14.0	9,829	934	10.5	15,167	2,722	5.6	9,627	790	12.2
1976	[4]9,418	[4]681	[4]13.8	10,127	934	10.8	15,438	2,764	5.6	9,774	806	12.1
1977	[4]9,517	[4]676	[4]14.1	10,607	947	11.2	16,700	3,002	5.6	9,978	814	12.3
1978	[4]9,500	[4]665	[4]14.3	10,968	948	11.6	18,045	3,263	5.5	10,077	816	12.4
1979	[4]9,062	[4]620	[4]14.6	10,802	905	11.9	18,502	3,380	5.5	9,722	776	12.5
1980	[4]8,813	[4]551	[4]16.0	10,437	854	12.2	18,736	3,447	5.4	9,458	712	13.3
1981	[4]8,873	[4]538	[4]16.5	10,244	819	12.5	19,016	3,565	5.3	9,477	697	13.6
1982	[4]9,050	[4]535	[4]16.9	10,276	762	13.5	19,931	3,647	5.5	9,644	686	14.1
1983	[4]9,118	[4]534	[4]17.1	10,497	767	13.7	21,083	3,769	5.6	9,760	686	14.2

TABLE 9.4

Motor vehicle mileage, fuel consumption, and fuel rates, 1949–1999 [CONTINUED]

Year	Passenger cars			Vans, pickup trucks, and sport utility vehicles[1]			Trucks[2]			All motor vehicles[3]		
	Mileage (miles per vehicle)	Fuel consumption (gallons per vehicle)	Fuel rate (miles per gallon)	Mileage (miles per vehicle)	Fuel consumption (gallons per vehicle)	Fuel rate (miles per gallon)	Mileage (miles per vehicle)	Fuel consumption (gallons per vehicle)	Fuel Rate (miles per gallon)	Mileage (miles per vehicle)	Fuel consumption (gallons per vehicle)	Fuel rate (miles per gallon)
1984	[4]9,248	[4]530	[4]17.4	11,151	797	14.0	22,550	3,967	5.7	10,017	691	14.5
1985	[4]9,419	[4]538	[4]17.5	10,506	735	14.3	20,597	3,570	5.8	10,020	685	14.6
1986	[4]9,464	[4]543	[4]17.4	10,764	738	14.6	22,143	3,821	5.8	10,143	692	14.7
1987	[4]9,720	[4]539	[4]18.0	11,114	744	14.9	23,349	3,937	5.9	10,453	694	15.1
1988	[4]9,972	[4]531	[4]18.8	11,465	745	15.4	22,485	3,736	6.0	10,721	688	15.6
1989	[4]10,157	[4]533	[4]19.0	11,676	724	16.1	22,926	3,776	6.1	10,932	688	15.9
1990	10,504	520	20.2	11,902	738	16.1	23,603	3,953	6.0	11,107	677	16.4
1991	10,571	501	21.1	12,245	721	17.0	24,229	4,047	6.0	11,294	669	16.9
1992	10,857	517	21.0	12,381	717	17.3	25,373	4,210	6.0	11,558	683	16.9
1993	10,804	527	20.5	12,430	714	17.4	26,262	4,309	6.1	11,595	693	16.7
1994	10,992	531	20.7	12,156	701	17.3	25,838	4,202	6.1	11,683	698	16.7
1995	11,203	530	21.1	12,018	694	17.3	26,514	4,315	6.1	11,793	700	16.8
1996	11,330	534	21.2	11,811	685	17.2	26,092	4,221	6.2	11,813	700	16.9
1997	11,581	539	21.5	12,115	703	17.2	27,032	4,218	6.4	12,107	711	17.0
1998	R11,754	R544	R21.6	R12,173	R707	R17.2	R25,397	R4,135	R6.1	R12,211	R721	R16.9
1999P	11,850	552	21.4	11,958	700	17.1	26,015	4,282	6.1	12,208	729	16.8

[1]Includes a small number of trucks with 2 axles and 4 tires, such as step vans.
[2]Single-unit trucks with 2 axles and 6 or more tires, and combination trucks.
[3]Includes buses and motorcycles, which are not shown separately.
[4]Includes motorcycles.
[5]Included in "Trucks."
[6]Includes vans, pickup trucks, and sport utility vehicles.
R=Revised. P=Preliminary.

SOURCE: "Table 2.8. Motor Vehicle Mileage, Fuel Consumption, and Fuel Rates, 1949–1999," in *Annual Energy Review 2000*, U.S. Department of Energy, Energy Information Administration, Washington, DC, 2001

TABLE 9.5

Alternative fuel vehicle acquisition mandates for centrally fueled fleets of federal agencies, state governments, and alternative fuel providers, 1996–2001 and beyond

Year	Percentage of all acquisitions for groups mandated to acquire vehicles		
	Federal agencies	State governments	Alternative fuel providers
1996	25	N/A	N/A
1997	33	10	30
1998	50	15	50
1999	75	25	70
2000	75	50	90
2001 and beyond	75	75	90

Note: The act mandated that the federal government had to acquire 5,000 alternative fuel vehicles in 1993; 7,500 vehicles in 1994; and 10,000 vehicles in 1995. It did not require state governments and alternative fuel providers to acquire alternative fuel vehicles during these years. In addition, the states' and fuel providers' acquisition mandates for 1996 were postponed for 1 year.

SOURCE: "Table 1. Alternative fuel vehicle acquisition mandates for centrally fueled fleets of federal agencies, state governments, and alternative fuel providers," in *Energy Policy Act of 1992: Limited Progress in Acquiring Alternative Fuel Vehicles and Reaching Fuel Goals,* U.S. General Accounting Office, Washington, DC, February 2000

Automakers and buyers of trucks and SUVs have opposed tightening emissions for these vehicles, although critics contend that new SUVs are more like cars than trucks in design and should fall under the same rules. As of November 2002 no new restrictions had been enacted, and these vehicles could still by law emit three times more nitrogen oxides, the main cause of smog, than cars.

The potential for savings from increased fuel economy in large trucks is huge, since their current fuel economy is so much lower than that of automobiles. In 2001 the EIA projected a small increase in fuel efficiency for the heavy-truck fleet and a larger (but still small) increase for the small-truck fleet. If the heavy-truck fleet were to reach a fuel efficiency of 10 mpg through technological improvements, projected oil demand would drop by 300,000 barrels per day. Over the past few years, aerodynamically designed trucks have become more common on American roads.

Cheap gasoline prices throughout the 1990s took away the sense of urgency surrounding fuel efficiency, which was demonstrated by the high growth of large vehicle sales. In addition, when the federal 55-mph speed limit law was repealed, many states allowed increased speed limits, which also lowered fuel efficiency. Carmakers have resisted building highly efficient cars, claiming that government mandates would saddle American motorists with car features they would not want and might not buy. In contrast, the European Commission has proposed an ambitious target of 47 mpg for gasoline-driven cars (compared to the current average of 29 mpg) and 52 mpg for diesel-powered cars by 2005. The general secretary of the European Council of Automotive Research and Development also announced in January 2001 that all 10 European auto manufacturers plan to build cars with low carbon dioxide emissions and high mileage by 2008.

It must be noted that while European countries do not generally legislate fuel efficiency, the cost of gasoline in Europe is more than twice that in the United States. That serves as a powerful incentive to European drivers to buy fuel-efficient vehicles. Many experts also see Europe as having a history of energy consumption not matched by the United States.

Alternative Fuel Vehicles (AFVs)

MANDATING AFVS. In recent years several laws have been passed to encourage or mandate the use of vehicles powered by fuels other than gasoline. The Clean Air Act Amendments of 1990 (PL101-549) required certain businesses and local governments with fleets of 10 or more vehicles in 21 metropolitan areas nationwide to phase in alternative fuel vehicles (AFVs) over time—20 percent of those fleets had to be AFVs by 1998. While great strides have been made in increasing the use of AFVs, there is no way to determine current compliance with the mandates because reporting and enforcement methods are inadequate.

The Energy Policy Act of 1992 (PL 102-486) was passed in the wake of the 1991 Persian Gulf War to conserve energy and increase the proportion of energy supplied domestically. It required the federal government to purchase 22,500 AFVs by 1995 and increase the percentage of AFV acquisitions from 25 percent of all acquisitions in 1996 to 75 percent in 1999 and thereafter. (See Table 9.5.) Agency budget cuts and inadequate enforcement have slowed compliance with these regulations. Still, many municipal governments and the U.S. Postal Service have put into operation fleets of natural gas vehicles, such as garbage trucks, transit buses, and postal vans.

NUMBERS AND TYPES OF AFVS. In 1993, 314,848 AFVs were on U.S. roads. Projections for 2002 expected 518,919 AFVs to be in use. (See Table 9.6.) These totals include vehicles originally manufactured to run on alternative fuels, as well as converted gasoline or diesel vehicles. The manufacture of new AFVs has been steadily increasing.

A number of different types of fuels are used in AFVs:

- Liquefied petroleum gas (LPG) is a mixture of propane and butane. Most AFVs in use in 2002 (54 percent) used LPG.

- Compressed natural gas (CNG) is natural gas that is stored in pressurized tanks. CNG releases one-tenth the carbon monoxide, hydrocarbon, and nitrogen of gasoline. It is the second most common AFV fuel, powering 24 percent of all AFVs in 2002.

- Methanol is a liquid fuel that can be produced from natural gas, coal, or biomass (plant material, vegetation, or agricultural waste).

TABLE 9.6

Estimated number of alternative-fueled vehicles in use, by fuel, 1993–2002

Fuel	1993	1994	1995	1996	1997	1998	1999	2000	2001	2002
Liquefied Petroleum Gases(LPG)	269,000	264,000	259,000	263,000	263,000	266,000	267,833	272,193	276,597	281,286
Compressed Natural Gas (CNG)	32,714	41,227	50,218	60,144	68,571	78,782	91,267	100,738	113,835	126,341
Liquefied Natural Gas (LNG)	299	484	603	663	813	1,172	1,681	2,090	2,576	3,187
Methanol, 85 Percent (M85)[a]	10,263	15,484	18,319	20,265	21,040	19,648	18,964	10,426	7,827	5,873
Methanol, Neat (M100)[a]	414	415	386	172	172	200	198	0	0	0
Ethanol, 85 Percent (E85)[a,b]	441	605	1,527	4,536	9,130	12,788	24,604	58,621	71,336	82,477
Ethanol, 95 Percent (E95)[a]	27	33	136	361	347	14	14	4	0	0
Electricity	1,690	2,224	2,860	3,280	4,453	5,243	6,964	11,834	17,848	19,755
Non-LPG Subtotal	**45,848**	**60,472**	**74,049**	**89,421**	**104,526**	**117,847**	**143,692**	**183,713**	**213,422**	**237,633**
Total	**314,848**	**324,472**	**333,049**	**352,421**	**367,526**	**383,847**	**411,525**	**455,906**	**490,019**	**518,919**

[a]The remaining portion of 85-percent methanol and both ethanol fuels is gasoline.
[b]In 1997, some vehicle manufacturers began including E85-fueling capability in certain model lines of vehicles. For 2000, the EIA estimated that the number of E-85 vehicles that are capable of operating on E85, gasoline, or both, is 2,652,592. Many of these AFVs are sold and used as traditional gasoline-powered vehicles. In this table, alternative fuel vehicles (AFV's) in use include only those E85 vehicles believed to be intended for use as alternative-fuel vehicles (AFVs). These are primarily fleet-operated vehicles.
Notes: Estimates for 2000 are revised. Estimates for 2001 are preliminary and estimates for 2002, in italics, are based on plans or projections. Estimates for historical years may be revised in future reports if new information becomes available. Beginning in 1999, LPG estimates are no longer rounded to the nearest thousand. Previously, the estimates were rounded in order to reflect the greatest uncertainty. Vehicle counts reported by the federal Agencies for years 2000, 2001, and 2002, may include vehicles ordered but not delivered. The greatest discrepancy between reported and actual in use vehicles occurs for the U.S. Postal Service.

SOURCE: "Table 1. Estimated Number of Alternative-Fueled Vehicles in Use in the United States, by Fuel, 1993-2002," in *Alternatives to Traditional Transportation Fuels 2000*, U.S. Department of Energy, Energy Information Administration, Washington, DC, September 2002 [Online] http://www.eia.doe.gov/cneaf/alternate/page/datatables/table1.html [accessed September 22, 2002]

• Ethanol is ethyl alcohol, a grain alcohol, mixed with gasoline and sold as gasohol.

• Liquefied natural gas (LNG) is natural gas (mostly methane) that has been liquefied by reducing its temperature to -260 degrees Fahrenheit.

• Electricity can be used for battery-powered, fuel cell, or hybrid vehicles.

• Biodiesels are liquid biofuels made from soybean, rapeseed, or sunflower oil, or from animal tallow. They can also be made from agricultural products, such as rice hulls.

The largest numbers of AFVs are located in California, Texas, Illinois, Ohio, and Michigan—together, these five states account for 38 percent of the projected U.S. total of AFVs in 2002. (See Table 9.7.) Transit buses are one type of heavy-duty vehicle that has seen much AFV activity. In 1999 approximately one in four new transit buses on order had alternative fuel capabilities.

ALTERNATIVE FUEL AND THE MARKETPLACE. AFVs cannot become a viable transportation option unless a fuel supply is readily available. Ideally, an infrastructure for supplying alternative fuels would be developed simultaneously with the AFVs. Table 9.8 shows the types and numbers of alternative fuel stations available in each state. There were 5,655 alternative refueling sites in the United States as of September 22, 2002. In 2002 privately owned vehicles used 73 percent of alternative fuel, state and local vehicles used 24 percent, and federal vehicles used 3 percent. (See Table 9.9.)

Many state policies and programs encourage the use of alternative fuels. California, for example, requires that 10 percent of vehicles for sale in the state by 2003 be zero-emission vehicles, such as hybrid electric vehicles. This has caused vehicle manufacturers to expedite vehicle research and development. In fact, electric vehicles are already selling in California, and some rental car agencies now offer them to customers at prices only slightly higher than those for gasoline-powered cars.

Chrysler Corporation stopped making natural gas–powered vehicles after the 1997 model year because it had lost money on the vehicles, selling only 4,000 after production began in 1992. General Motors, which had suspended sales of natural-gas vehicles in 1994, resumed sales in 1997. Ford began selling some natural-gas versions of its cars and trucks in 1995. Commercial fleets, not retail customers, are the main buyers of natural-gas vehicles.

Market success of alternative fuels and AFVs depends upon public acceptance. People are accustomed to using gasoline as their main transportation fuel and it is readily available. As federal and state requirements for alternative fuels increase, so will the fuels' visibility and acceptance by the general public.

ELECTRIC CARS: PROMISE AND REALITY. In the early days of the automobile, electric cars outnumbered internal-combustion vehicles. With the introduction of technology for producing low-cost gasoline, however, electric vehicles fell out of favor. But as cities became choked with air pollution, the idea of an efficient electric car emerged. To make it acceptable to the public, however, several

TABLE 9.7

Estimated number of alternative-fueled vehicles in use, by state and fuel type, 2000

State	Liquefied Petroleum Gases	Natural Gas	Methanol	Ethanol	Electricity	Total
Alabama	2,515	583	0	1,042	84	4,224
Alaska	246	93	0	0	4	343
Arizona	2,579	5,664	357	591	1173	10,364
Arkansas	907	926	0	74	5	1,912
California	33,689	19,128	8,509	2,374	3856	67,556
Colorado	4,952	4,382	5	1,795	301	11,435
Connecticut	1,479	1,106	1	648	52	3,286
Delaware	270	340	17	293	3	923
District of Columbia	35	548	88	2,080	163	2,914
Florida	8,501	3,198	11	1,007	634	13,351
Georgia	8,290	3,890	69	987	1073	14,309
Hawaii	410	0	0	594	178	1,182
Idaho	1,474	396	0	931	108	2,909
Illinois	15,900	2,197	31	4,210	163	22,501
Indiana	7,707	2,556	0	2,345	52	12,660
Iowa	5,627	295	48	1,899	14	7,883
Kansas	1,438	52	1	1,416	6	2,913
Kentucky	3,371	581	0	2,977	24	6,953
Louisiana	3,252	723	5	691	30	4,701
Maine	491	3	0	58	18	570
Maryland	3,343	1,390	12	1,364	9	6,118
Massachusetts	3,091	930	64	726	506	5,317
Michigan	14,235	1,905	85	2,798	483	19,506
Minnesota	2,133	525	0	3,598	11	6,267
Mississippi	5,157	134	0	108	4	5,403
Missouri	3,504	843	168	2,026	49	6,590
Montana	1,380	391	0	458	71	2,300
Nebraska	2,687	321	0	1,693	6	4,707
Nevada	1,412	3,456	0	256	32	5,156
New Hampshire	473	12	0	52	52	589
New Jersey	4,091	2,211	6	808	146	7,262
New Mexico	3,607	1,070	18	829	89	5,613
New York	8,262	6,846	156	411	810	16,485
North Carolina	9,052	391	0	1,674	54	11,171
North Dakota	555	500	0	689	20	1,764
Ohio	15,119	3,954	46	2,896	122	22,137
Oklahoma	10,158	5,784	0	1,074	160	17,176
Oregon	7,232	496	34	223	57	8,042
Pennsylvania	10,133	3,083	192	1,252	81	14,741
Puerto Rico	0	0	0	0	68	68
Rhode Island	487	281	0	46	18	832
South Carolina	3,771	146	0	907	23	4,847
South Dakota	878	99	0	628	9	1,614
Tennessee	8,504	577	0	1,126	55	10,262
Texas	29,989	9,959	287	2,374	234	42,843
Utah	1,756	3,552	14	657	33	6,012
Vermont	313	7	0	89	24	433
Virginia	3,952	1,972	12	747	225	6,908
Washington	5,002	2,124	129	323	326	7,904
West Virginia	665	1,615	0	143	5	2,428
Wisconsin	6,988	1,521	61	2,625	41	11,236
Wyoming	1,131	72	0	13	70	1,286
U.S. Total	**272,193**	**102,828**	**10,426**	**58,625**	**11,834**	**455,906**

Notes: Natural gas includes compressed natural gas (CNG) and liquefied natural gas (LNG). Methanol includes M85 and M100. Ethanol includes E85 and E95. Totals may not equal sum of components due to independent rounding. Beginning in 1999, LPG estimates are no longer rounded to the nearest thousand. Previously, the estimates were rounded in order to reflect the greatest uncertainty. Vehicle counts reported by the federal agencies for years 2000, 2001, and 2002, may include vehicles ordered but not delivered. The greatest discrepancy between reported and actual in use vehicles occurs for the U.S. Postal Service.

SOURCE: "Table 4. Estimated Number of Alternative-Fueled Vehicles in Use, by State and Fuel Type, 2000," in *Alternatives to Traditional Transportation Fuels 2000,* U.S. Department of Energy, Energy Information Administration, Washington, DC, September 2002 [Online] http://www.eia.doe.gov/cneaf/alternate/page/datatables/table4.html [accessed September 22, 2002]

TABLE 9.8

Alternative fuel station counts, by state and fuel type, as of September 22, 2002

State	CNG	E85	LPG	ELEC	BD	LNG	All
Alabama	14		67	34		2	117
Alaska	0	0	8	0	0	0	8
Arizona	30	1	108	63	2	3	207
Arkansas	7	0	78	0	0	0	85
California	215	0	336	540	8	9	1108
Colorado	39	7	80	6	0	1	133
Connecticut	25	0	29	0	0	0	54
Delaware	4	0	4	0	0	0	8
Dist. of Columbia	3	0	0	0	0	0	3
Florida	40	0	144	3	0	1	188
Georgia	67	0	53	82	0	2	204
Hawaii	0	0	7	11	1	0	19
Idaho	8	1	33	0	0	0	42
Illinois	22	15	81	0	0	0	118
Indiana	32	1	48	0	1	3	85
Iowa	0	11	42	0	0	0	53
Kansas	5	1	67	0	0	1	74
Kentucky	6	7	23	0	0	0	36
Louisiana	14	0	41	0	0	0	55
Maine	0	0	20	0	1	0	21
Maryland	30	1	28	1	0	1	61
Massachusetts	12	0	38	25	1	0	76
Michigan	23	8	125	0	2	1	159
Minnesota	11	67	56	0	0	1	135
Mississippi	3	0	32	0	0	0	35
Missouri	7	5	149	0	1	0	162
Montana	9	2	40	0	0	1	52
Nebraska	5	7	27	0	0	0	39
Nevada	20	0	34	0	1	0	55
New Hampshire	1	0	30	12	0	0	43
New Jersey	30	0	29	0	0	0	59
New Mexico	15	1	81	0	0	1	98
New York	60	0	95	16	0	0	171
North Carolina	10	0	75	6	0	0	91
North Dakota	4	2	18	0	0	0	24
Ohio	52	2	73	0	1	1	129
Oklahoma	58	0	92	0	0	0	150
Oregon	16	0	49	4	2	1	72
Pennsylvania	55	0	104	0	0	1	160
Rhode Island	6	0	7	0	0	0	13
South Carolina	4	1	62	0	1	0	68
South Dakota	2	8	26	0	0	0	36
Tennessee	2	0	59	0	0	0	61
Texas	66	0	434	7	1	7	515
Utah	62	2	36	0	0	1	101
Vermont	0	0	17	11	0	0	28
Virginia	24	1	64	11	1	3	104
Washington	25	0	89	6	1	1	122
West Virginia	43	0	10	0	0	0	53
Wisconsin	22	3	84	0	0	0	109
Wyoming	18	0	37	0	0	1	56
TOTALS:	**1226**	**154**	**3369**	**838**	**25**	**43**	**5655**

Note: CNG=Compressed Natural Gas
E85=85% Ethanol
LPG=Propane
ELEC=Electric
BD=Bio Diesel
LNG=Liquified Natural Gas

SOURCE: "Alternative Fuel Station Counts listed by State and Fuel Type," in *Refueling Stations,* Alternative Fuels Data Center, 2002 [Online] http://www.afdc.nrel.gov/refuel/state_tot.shtml [accessed September 22, 2002]

TABLE 9.9

Estimated consumption of alternative transportation fuels, by vehicle ownership, 1998, 2000, and 2002

Fuel	1998				2000				2002			
	Federal	State and Local	Private	Total	Federal	State and Local	Private	Total	Federal	State and Local	Private	Total
Liquefied Petroleum Gases (LPG)	71	26,384	214,931	**241,386**	474	27,381	219,207	**247,062**	*545*	*28,192*	*226,778*	***255,515***
Compressed Natural Gas (CNG)	4,572	33,627	34,213	**72,412**	6,286	50,113	41,952	**98,351**	*7,927*	*52,905*	*52,722*	***113,554***
Liquefied Natural Gas (LNG)	125	4,294	924	**5,343**	189	5,343	1,589	**7,121**	*259*	*7,910*	*2,335*	***10,504***
Methanol, 85 Percent (M85)[a]	13	325	874	**1,212**	2	193	390	**585**	*0*	*109*	*221*	***330***
Methanol, Neat (M100)	0	449	0	**449**	0	437	0	**437**	*0*	*0*	*0*	***0***
Ethanol, 85 Percent(E85)[a]	425	794	508	**1,727**	2,659	2,851	1,564	**7,074**	*3,315*	*3,817*	*2,943*	***10,075***
Ethanol, 95 Percent (E95)[a]	0	59	0	**59**	0	13	0	**13**	*0*	*0*	*0*	***0***
Electricity	46	471	685	**1,202**	650	1,122	898	**2,670**	*991*	*1,941*	*1,528*	***4,460***
Total	**5,252**	**66,403**	**252,135**	**323,790**	**10,260**	**87,453**	**265,600**	**363,313**	***13,037***	***94,874***	***286,527***	***394,438***

[a]The remaining portion of 85-percent methanol and both ethanol fuels is gasoline. Consumption data include the gasoline portion of the fuel.

Notes: Fuel quantities are expressed in a common base unit of gasoline-equivalent gallons to allow comparisons of different fuel types. Gasoline-equivalent gallons do not represent gasoline displacement. Gasoline equivalent is computed by dividing the lower heating value of the alternative fuel by the lower heating value of gasoline and multiplying this result by the alternative fuel consumption value. Lower heating value refers to the Btu content per unit of fuel excluding the heat produced by condensation of water vapor in the fuel. Totals may not equal sum of components due to independent rounding. Estimates for 2002, in italics, are based on plans or projections. Estimates for historical years may be revised in future reports if new information becomes available.

SOURCE: "Table 13. Estimated Consumption of Alternative Transportation Fuels in the United States, by Vehicle Ownership, 1998, 2000, and 2002 (Thousand Gasoline-Equivalent Gallons)" in *Alternatives to Traditional Transportation Fuels 2000*, U.S. Department of Energy, Energy Information Administration, Washington, DC, September 2002 [Online] http://www.eia.doe.gov/cneaf/alternate/page/datatables/table13.html [accessed September 22, 2002]

considerations had to be addressed: How many miles could an electric car be driven before needing to be recharged? How light would the vehicle need to be? And could the electric car keep up with the speed and driving conditions of busy freeways and highways?

In 1993 tax breaks became available for people who buy cars that run on alternative energy sources; these breaks were especially generous for purchasers of electric cars. The breaks were intended to compensate for the price difference between electric cars and the average gas-powered car, and to jump-start production of these vehicles. By 2003, 10 percent of all new cars offered for sale in California must be zero-emission vehicles. New York, Massachusetts, Maine, and Vermont each have similar laws all set to take effect in 2003.

Electric vehicles (EVs) come in three types: battery-powered; fuel cell; and hybrids, which are powered by both an electric motor and a small conventional engine. EV1, a two-seater by General Motors (GM), was the first commercially available electric car. In 1999 GM introduced its second-generation EV1, the Gen II. It uses a lead-acid battery pack and has a driving range of approximately 95 miles. The Gen II is also offered with an optional nickel-metal hydride battery pack, which increases its range to 130 miles. Ford is currently producing a Ford Ranger in an EV model. EV drivers have a charger installed at their home, allowing them to recharge the car overnight. Rechargers are also available in some public places.

Fuel cell electric vehicles use an electrochemical process that converts a fuel's energy into usable electricity. Some experts think that in the future vehicles driven by

fuel cells could replace vehicles with combustion engines. Fuel cells produce very little sulfur and nitrogen dioxide and generate less than half the carbon dioxide of internal-combustion engines. Rather than needing to be recharged, they are simply refueled. Hydrogen, natural gas, methanol, and gasoline can all be used with a fuel cell. The Bush administration and Secretary of Energy Spencer Abraham expressed interest in promoting fuel-cell technology.

DaimlerChrysler's Mercedes-Benz division produced the first prototype fuel-cell car. The NECAR4 produces zero emissions and runs on liquid hydrogen. The hydrogen must be kept cold at all times, which makes the design impractical for widespread use. However, the company plans to replace the NECAR4 with the NECARX, which will run on methanol and is expected to be more practical. The NECAR4 prototype travels 280 miles on a full 11-gallon tank. It was unveiled in 1999, and consumers may be able to purchase a production NECARX by 2004.

Ecostar, an alliance between Ford, DaimlerChrysler, and Ballard Power Systems, is working on developing new fuel cells to power vehicles. Other automakers have experimented with fuel-cell prototype cars as well, but these vehicles are not yet commercially available. They are expected to be on the roads by 2004.

Hybrid cars have both an electric motor and a small internal-combustion engine. A sophisticated computer system automatically shifts from the electric motor to the gas engine, as needed, for optimum driving. The electric motor is recharged while the car is driving and braking. Because the gasoline engine does only part of the work,

FIGURE 9.5

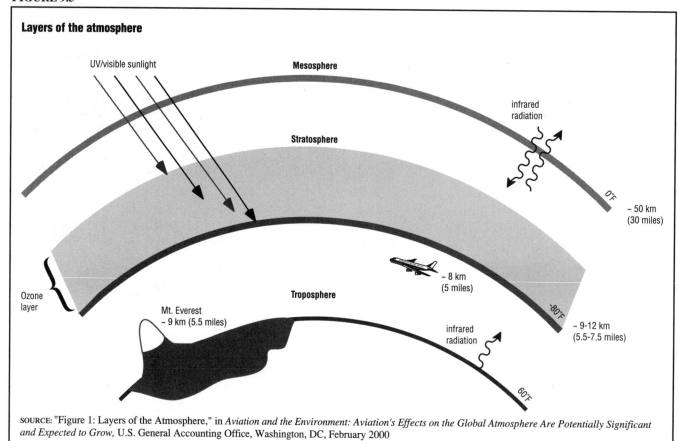

Layers of the atmosphere

UV/visible sunlight

Mesosphere

Stratosphere

infrared radiation

~ 50 km (30 miles)

0°F

~ 8 km (5 miles)

Ozone layer

Troposphere

Mt. Everest ~ 9 km (5.5 miles)

infrared radiation

-80°F

~ 9-12 km (5.5-7.5 miles)

60°F

SOURCE: "Figure 1: Layers of the Atmosphere," in *Aviation and the Environment: Aviation's Effects on the Global Atmosphere Are Potentially Significant and Expected to Grow,* U.S. General Accounting Office, Washington, DC, February 2000

these cars get very good fuel economy. The engines are also designed for ultralow emissions.

As of 2002 there were two hybrids on the market in the United States: the Toyota Prius, a comfortable sedan with front and back seating, and the two-passenger Honda Insight. Both cars were sold in Japan for several years before being introduced to the U.S. market. Ford created a hybrid SUV, the Ford Escape, planned to be available in 2003.

Air Travel Efficiency

The average American flew 1,739 miles in 2000. Europeans, though they flew fewer miles, had the world's most crowded skies, while the most rapid growth in flying was in Asia. Most air travel is done by a small portion of the world's people.

Flying carries an environmental price: It is a very energy-intensive form of transportation. In much of the industrialized world, air travel is replacing more energy-efficient rail or bus travel. The U.S. Department of Transportation reports that despite a rise in the fuel efficiency of jet engines, jet fuel consumption rose 20 percent between 1995 and 2000 and is expected to double from 1995 levels by 2012.

Another problem with air travel is its impact on global warming. Airplanes spew nearly 4 million tons of nitro-

gen oxide into the air, much of it while cruising in the tropospheric zone five to seven miles above the Earth where ozone is formed. (See Figure 9.5.) The EPA estimates that air traffic accounts for about 3 percent of all global greenhouse warming. The IPCC notes that emissions deposited directly into the atmosphere do greater harm than those released at the Earth's surface.

In 1996 Pratt and Whitney, a designer and manufacturer of high-performance engines, announced plans to introduce a radical new engine design that would be cleaner, more efficient, quieter, and more reliable than conventional designs. The new engine was undergoing detailed design in 2002, will be tested in 2004, and is expected to take its first flight in 2006. The engine would reduce emissions by 40 percent and exceed noise restrictions that took effect in 2000. The engine is designed for use on single-aisle planes carrying 120 to 180 passengers, such as the Airbus A320.

Although each generation of airplane engine gets cleaner and more fuel-efficient, there are also other engines in the airline industry—those in the trucks, cars, and carts that service airplane fleets. Electric utility companies, including the Edison Electric Institute and the Electric Power Research Institute, launched a program in 1993 to electrify airports. By converting terminal transport buses, food trucks, and baggage-handling carts to electricity, airports could reduce air pollution considerably. As of December 2002, only a few U.S. airports and airlines were operating significant numbers of electric ground support equipment and the associated electric charging stations. However, airport electrificiation implementation and research were ongoing.

CONSERVATION IN THE RESIDENTIAL AND COMMERCIAL SECTORS

Total building energy use in the United States has increased: There are increasing numbers of people, households, and offices. However, energy use per unit area (commercial) or per person (residential) has roughly stabilized over the past 10 to 12 years because of a variety of efficiency improvements. The sources of energy in buildings have changed dramatically. Use of fuel oil has dropped, and natural gas has largely made up the difference. At the same time, other energy demands have risen. Electronic office equipment, such as computers, fax machines, printers, and copiers, has sharply increased electricity loads in commercial buildings. Energy use in buildings accounts for an increasing share of total U.S. energy consumption: 27 percent in 1950, 33 percent in 1970, and 35 percent in 1999. The residential and commercial sectors used roughly 38 percent of U.S. energy in 2000. (See Figure 1.10 in Chapter 1.)

Building Efficiency

There are several potential areas for research and development in energy conservation in buildings. Among the techniques useful in reducing energy loads are advanced window designs, daylighting (letting light in from the outside by using high windows, skylights, and atria in the center of large buildings), solar water heating, landscaping, and tree planting. Energy conservation efforts in the building sector have been substantial since the early 1980s.

Residential energy consumption can be reduced by introducing more efficient new housing and appliances, improving the existing stock of housing, and by building more multiple-family units. Residential energy consumption is reduced as well when people migrate to the South and West, where the combined use of heating and cooling is generally lower than in other parts of the country.

In the residential sector, the amount of energy used in newer homes, particularly those built since 1980, is dramatically less than that used in older homes. The largest share of energy savings is the result of better construction, higher quality insulation, and more energy-efficient windows and doors. The Office of Technology Assessment reported in May 1992 (in *Building Energy Efficiency*) that roughly one-fourth the energy used to heat and cool buildings is lost through poor insulation and poorly insulated windows. Before the 1973 energy crisis, 70 percent of

FIGURE 9.6

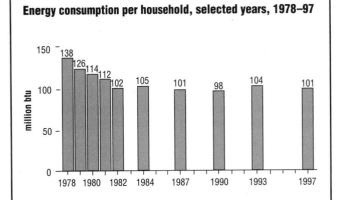

Energy consumption per household, selected years, 1978–97

Notes: No data are available for years not shown. Data for 1978 through 1984 are April of the year shown through March of the following year; data for 1987, 1990, 1993, and 1997 are for the calendar year.

SOURCE: Adapted from "Figure 2.4. Household Energy Consumption: Consumption per Household, Selected years, 1978–1997," in *Annual Energy Review 2000*, U.S. Department of Energy, Energy Information Administration, Washington, DC, 2001.

new windows sold were single-glazed. By 1990, because of changes in building codes and public interest, 80 percent of windows sold were double-glazed with double insulating ability, cutting energy loss in half.

Energy consumption per household has remained fairly steady since 1982, as technology gains have been offset by an increase in the size of new homes and more demand for energy services. (See Figure 9.6.) As in the residential sector, improved technology has helped to slow the growth in commercial building energy use. Commercial buildings constructed after 1980 use considerably less energy than those built in the early part of the 1900s.

Home Appliance Efficiency

Overall, the number of households in the United States is increasing, which is increasing the demand for energy-intensive services like air-conditioning and appliances. The EIA reports that residential energy use accounted for about 21 percent of the total national energy use in 2000. (See Figure 1.10 in Chapter 1.) For household energy consumption in 1997 (the most recent data available), space heating used 51 percent of the total energy consumed; appliances, 26 percent; water heating, 19 percent; and air conditioners, 4 percent.

The number of appliances in households has been increasing steadily over the past several decades. (See Figure 9.7.) By 1997, 99 percent of American homes had color televisions, 47 percent had central air-conditioning, and 35 percent had personal computers.

In 1987 Congress passed the National Appliance Energy Conservation Act (NAECA; PL 95-629), which gave the DOE the authority to formulate minimum

FIGURE 9.7

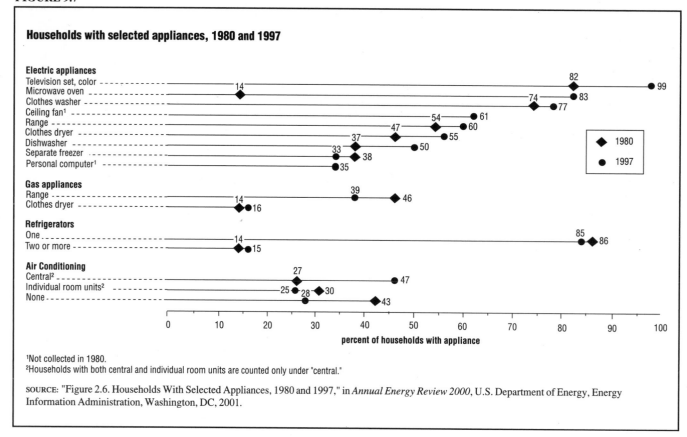

Households with selected appliances, 1980 and 1997

Electric appliances
Television set, color
Microwave oven
Clothes washer
Ceiling fan[1]
Range
Clothes dryer
Dishwasher
Separate freezer
Personal computer[1]

Gas appliances
Range
Clothes dryer

Refrigerators
One
Two or more

Air Conditioning
Central[2]
Individual room units[2]
None

percent of households with appliance

◆ 1980
● 1997

[1]Not collected in 1980.
[2]Households with both central and individual room units are counted only under "central."

SOURCE: "Figure 2.6. Households With Selected Appliances, 1980 and 1997," in *Annual Energy Review 2000*, U.S. Department of Energy, Energy Information Administration, Washington, DC, 2001.

efficiency requirements for 13 classes of consumer products. It could also revise and update those standards as technologies and economic conditions changed. Table 9.10 shows the products affected and the years in which appliance efficiency standards were established or revised for each, as well as the future effective dates of standards. In 1997 the DOE established an advisory committee to review and revise the standards.

Energy efficiency has increased for all major household appliances but most dramatically for refrigerators and freezers. Since 1972 new refrigerators and freezers have almost doubled in energy efficiency because of better insulation, motors, compressors, and accessories such as automatic defrost. These improvements have been accomplished at relatively low cost to manufacturers. In addition, efficiency labels for consumers are now required, which makes purchasing efficient models easier.

By the early 2000s, air conditioners and heat pumps, another major group of appliances, had shown a 30–50 percent improvement in energy efficiency since the mid-1970s. Although this improvement in energy efficiency was less than the improvements in refrigerators and freezers, it is significant because these appliances are large energy users.

Water heaters and furnaces improved efficiency between 5 percent and 20 percent from the mid-1970s to the early 2000s. However, the technological improvements in these appliances are relatively costly compared to the overall price of the product. This means that the more energy-conserving models have a higher retail price, which discourages many consumers from purchasing efficient models, even though the more efficient models may save money in the long run. In addition, many purchases of water heaters and furnaces are made by builders, who have little incentive to pay more for the most efficient models, or by homeowners in emergency situations, when fast availability and installation seem much more important than energy efficiency.

Nonetheless, consumers are sometimes willing to purchase more expensive, energy-efficient models of air conditioners, refrigerators, and lights if the devices can save them enough money in the long run on their electricity bills to offset the higher purchase costs. According to a U.S. General Accounting Office study (*Energy Conservation: Efforts Promoting More Efficient Use,* Washington, DC, 1992), consumers will purchase such devices if the "payback period" is two years or less.

In addition to concerns about efficiency, appliance makers, especially those who make refrigerators and air-conditioning systems, are striving to develop alternative cooling techniques as substitutes for chlorofluorocarbons (CFCs), which are ozone-damaging chemicals that can no longer be legally sold in the United States. Current technol-

TABLE 9.10

Effective dates of appliance efficiency standards, 1988–2007

Product	1988	1990	1992	1993	1994	1995	2000	2001	2003	2004	2005	2006	2007
Clothes dryers	X				X								
Clothes washers	X				X					X			X
Dishwashers	X				X								
Refrigerators and freezers		X		X				X					
Kitchen ranges and ovens		X											
Room air conditioners		X					X						
Direct heating equipment		X											
Fluorescent lamp ballasts		X									X		
Water heaters		X								X			
Pool heaters		X											
Central air conditioners and heat pumps			X									X	
Furnaces													
Central (>45,000 Btu per hour)			X										
Small (<45,000 Btu per hour)			X										
Mobile home		X											
Boilers			X										
Fluorescent lamps, 8 foot					X								
Fluorescent lamps, 2 and 4 foot (U tube)						X							
Commercial water-cooled air conditioners										X			
Commercial natural gas furnaces										X			
Commercial natural gas water heaters										X			

SOURCE: "Table 2. Effective dates of appliance efficiency standards, 1988–2007," in *Annual Energy Outlook 2002,* U.S. Department of Energy, Energy Information Administration, Washington, DC, 2001

ogy is temporarily substituting CFCs with somewhat less dangerous HCFCs (hydrochlorofluorocarbons). In Europe, refrigeration units using other substances, such as propane "greenfreeze" technology, are rapidly replacing HCFCs.

Lawn and Garden Equipment

In 1994 the EPA reported that as much as 10 percent of the nation's air pollution was generated by gasoline-powered lawn and garden equipment, including lawn mowers, chain saws, and golf carts. Former EPA administrator Carol Browner estimated that Americans use 89 million pieces of such equipment, with lawn mowers alone accounting for 5 percent of the nation's air pollution.

Under Browner, the agency established engine label and warranty requirements, exhaust emissions standards, and test procedures, requiring that engine makers meet the new requirements by 1996. Effective that year, new products offered for sale were equipped with improved carburetion systems, and additional standards were scheduled for subsequent years. Agency officials reported in 2000 that the new regulations had reduced smog-forming hydrocarbon emissions by 32 percent and that additional standards to be phased in between 2001 and 2007 are projected to reduce hydrocarbon emissions an additional 10 percent.

INTERNATIONAL COMPARISONS OF CONSERVATION EFFORTS

Compared to other industrialized countries, the United States is lagging behind in energy efficiency and conservation efforts. According to EIA figures for energy consumption per dollar of GDP in 2000, the United States consumed 10,570 Btu per dollar of GDP compared to 7,400 for France, 7,300 for Germany, and 6,600 for Japan in 1998. (See Figure 9.2.) This means other countries are consuming less energy per dollar of the output of goods and services; put simply, they are making products more efficiently. If fuel prices increase in the future, the United States may face economic challenges from more efficient countries.

The United States also falls behind other industrialized countries in controlling carbon emissions. As the world's largest producer of greenhouse gases, U.S. per capita emissions are also significantly higher than in other industrialized countries. For instance, EIA figures for carbon dioxide emissions from the consumption of fossil fuels in 1998 showed that the United States emitted 5.56 metric tons of carbon equivalent per person, compared to 2.8 metric tons per person in Germany, 2.3 metric tons per person in Japan, and 1.8 metric tons per person in France.

FUTURE TRENDS IN CONSERVATION

From the EIA

The *Annual Energy Outlook 2002* (EIA, 2001) expects U.S. total energy consumption to increase from 99.3 quadrillion Btu to 130.9 quadrillion Btu between 2000 and 2020, an average annual increase of 1.4 percent, even with efficiency standards for new energy-using equipment in buildings and for motors taken into consideration. Residential and industrial energy demands are expected to be below this average annual increase, at 1.0 percent per year and 1.1 percent per year, respectively. However, commercial and transportation energy demands

FIGURE 9.8

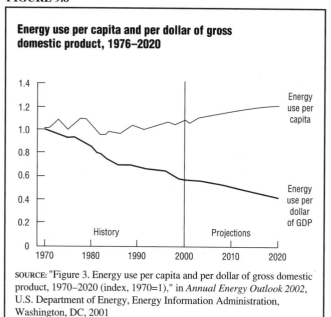

Energy use per capita and per dollar of gross domestic product, 1976–2020

SOURCE: "Figure 3. Energy use per capita and per dollar of gross domestic product, 1970–2020 (index, 1970=1)," in *Annual Energy Outlook 2002*, U.S. Department of Energy, Energy Information Administration, Washington, DC, 2001

FIGURE 9.9

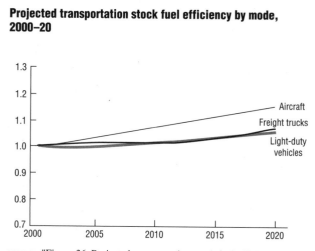

Projected transportation stock fuel efficiency by mode, 2000–20

SOURCE: "Figure 36. Projected transportation stock fuel efficiency by mode, 2000–2020 (index, 2000 = 1)," in *Annual Energy Outlook 2002*, U.S. Department of Energy, Energy Information Administration, Washington, DC, 2001

FIGURE 9.10

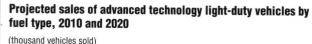

Projected sales of advanced technology light-duty vehicles by fuel type, 2010 and 2020

(thousand vehicles sold)

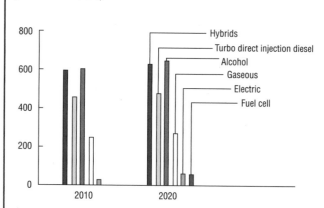

SOURCE: "Figure 38. Projected sales of advanced technology light-duty vehicles by fuel type, 2010 and 2020 (thousand vehicles sold)," in *Annual Energy Outlook 2002*, U.S. Department of Energy, Energy Information Administration, Washington, DC, 2001

will exceed the average, at 1.7 percent per year and 1.9 percent per year, respectively.

Per capita energy use is expected to increase slightly by 0.6 percent per year through 2020. (See Figure 9.8.) Although projections anticipate homes will be larger, electricity will be used more intensively, and annual personal highway and air travel will increase, efficiency improvements will offset much of the increase. Energy use per dollar of GDP is also projected to decrease as efficiency gains are made.

The EIA projects that transportation fuel efficiency will grow more slowly from 2000 through 2020 than it did in the 1980s. (See Figure 9.9.) Light-duty vehicle efficiency is projected to remain mainly steady, because low fuel prices and higher personal incomes may lower the incentive for efficiency. For light-duty vehicles, including cars, the EIA projects that any gains in efficiency will by offset by consumer preferences for larger, more powerful vehicles. By 2020 fuel efficiency for large trucks is expected to increase only slightly, at about 6 percent, while a larger 16 percent gain is expected for aircraft.

The EIA predicts that the market for AFVs will grow as a result of federal and state mandates. (See Figure 9.10.) By 2020 AFVs are projected to account for 2.1 million vehicle sales, or 12 percent of total vehicle sales. According to the *Annual Energy Outlook 2002,* alternative fuels could replace about 184,000 barrels of oil per day by 2020, or 2 percent of all light-vehicle consumption.

From the Bush Administration

The National Energy Policy Plan (NEPP) of the Department of Energy Organization Act of 1977 (PL 95-91) requires the president to submit to Congress a national energy policy plan every two years. This plan includes energy objectives and strategies, as well as projections of energy supply, demand, and prices. In May 2001 the Bush administration submitted the most recent plan, *National Energy Policy: Report of the National Energy Policy Development Group.*

The report focuses on long-term strategies to provide reliable energy and a clean environment. To do so, the report proposes modernizing the nation's infrastructure, which involves repairing and expanding its outdated network of electric generators, transmission lines, pipelines, and refineries. The report also proposes a modernization of conservation by increasing funding for renewable energy and energy efficiency research, creating tax incentives for purchasing hybrid vehicles, extending the "Energy Star" efficiency program, and reviewing the CAFE standards for transportation.

The report recognizes America's dependence on a narrow range of energy sources and supports research into clean coal technologies, the increased use of nuclear power, and the use of new technologies in oil and natural gas exploration. The report also proposes opening a small part of the Arctic National Wildlife Refuge to environmentally regulated exploration. Finally, the report promotes further improvements in emission control and the export of environmentally friendly technologies to other parts of the world.

IMPORTANT NAMES AND ADDRESSES

American Gas Association
400 N. Capitol St., NW
Washington, DC 20001
(202) 824-7000
FAX: (202) 824-7115
URL: http://www.aga.org

American Petroleum Institute
1220 L St., NW
Washington, DC 20005-4070
(202) 682-8000
FAX: (202) 962-4730
E-mail: pr@api.org
URL: http://www.api.org

American Wind Energy Association
122 C St., NW, Ste. 380
Washington, DC 20001
(202) 383-2500
FAX: (202) 383-2505
E-mail: windmail@awea.org
URL: http://www.awea.org

Committee on Resources
U.S. House of Representatives
1324 Longworth House Office Bldg.
Washington, DC 20515-6201
(202) 225-2761
E-mail: resources.committee@mail.house.gov
URL: http://resourcescommittee.house.gov

Council on Environmental Quality
722 Jackson Pl., NW
Washington, DC 20503
(202) 395-5750
FAX: (202) 456-6546
URL: http://www.whitehouse.gov/ceq

Edison Electric Institute
701 Pennsylvania Ave., NW
Washington, DC 20004-2696
(202) 508-5000
FAX: (202) 508-5759
URL: http://www.eei.org

Electric Power Research Institute
3412 Hillview Ave.
Palo Alto, CA 94304
E-mail: askepri@epri.com
URL: http://www.epri.com

Energy Information Administration
U.S. Department of Energy
1000 Independence Ave., SW
Washington, DC 20585
(202) 586-8800
FAX: (202) 586-0727
E-mail: infoctr@eia.doe.gov
URL: http://www.eia.doe.gov

Environmental Defense
257 Park Ave. South
New York, NY 10010
(212) 505-2100
FAX: (212) 505-2375
URL: http://www.environmentaldefense.org

Environmental Industry Associations
4301 Connecticut Ave., NW, Ste. 300
Washington, DC 20008
(202) 244-4700
FAX: (202) 966-4818
Toll-free: (800) 424-2869
E-mail: membership@envasns.org
URL: http://www.envasns.org

Friends of the Earth
1025 Vermont Ave., NW, Ste. 300
Washington, DC 20005
FAX: (202) 783-0444
Toll-free: (877) 843-8687
E-mail: foe@foe.org
URL: http://www.foe.org

Greenpeace USA
702 H St., NW, Ste. 300
Washington, DC 20001
(202) 462-1177
FAX: (202) 462-4507
Toll-free: (800) 326-0959
URL: http://www.greenpeaceusa.org

National Mining Association
101 Constitution Ave., NW, Ste. 500 East
Washington, DC 20001-2133
(202) 463-2600
FAX: (202) 463-2666
E-mail: thowe@nma.org
URL: http://www.nma.org

Natural Gas Supply Association
805 15th St., NW, Ste. 510
Washington, DC 20005
(202) 326-9300
FAX: (202) 326-9330
URL: http://www.ngsa.org

Natural Resources Defense Council
40 West 20th St.
New York, NY 10011
(212) 727-2700
FAX: (212) 727-1773
E-mail: nrdcinfo@nrdc.org
URL: http://www.nrdc.org

Nuclear Energy Institute
1776 I St., NW, Ste. 400
Washington, DC 20006-3708
(202) 739-8000
FAX: (202) 785-4019
E-mail: webmasterp@nei.org
URL: http://www.nei.org

Public Citizen
1600 20th St., NW
Washington, DC 20009
(202) 588-1000
FAX: (202) 588-7799
E-mail: CMEP@citizen.org
URL: http://www.citizen.org

Sierra Club
85 2nd St., 2nd Fl.
San Francisco, CA 94105
(415) 977-5500
FAX: (415) 977-5799
E-mail: information@sierraclub.org
URL: http://www.sierraclub.org

Solid Waste Association of North America
P.O. Box 7219
Silver Spring, MD 20907-7219
(301) 585-2898
FAX: (301) 589-7068
Toll-free: (800) 467-9262
E-mail: info@swana.org
URL: http://www.swana.org

Union of Concerned Scientists
2 Brattle Sq.
Cambridge, MA 02238-9105
(617) 547-5552
FAX: (617) 864-9405
E-mail: ucs@ucsusa.org
URL: http://www.ucsusa.org

U.S. Bureau of Land Management
1849 C St., Rm. 406-LS
Washington, DC 20240
(202) 452-5125
FAX: (202) 452-5124
URL: http://www.blm.gov

U.S. Department of Energy
1000 Independence Ave., SW
Washington, DC 20585
FAX: (202) 586-4403
Toll-free: (800) DIAL-DOE
URL: http://www.energy.gov

U.S. Environmental Protection Agency
Ariel Rios Bldg.
1200 Pennsylvania Ave., NW
Washington, DC 20460
(202) 260-2090
FAX: (202) 564-0279
E-mail: public-access@epa.gov
URL: http://www.epa.gov

U.S. Nuclear Regulatory Commission
Washington, DC 20555
(301) 415-8200
FAX: (301) 415-2395
Toll-free: (800) 368-5642
E-mail: opa@nrc.gov
URL: http://www.nrc.gov

U.S. Senate Committee on Energy and Natural Resources
364 Dirksen Senate Bldg.
Washington, DC 20510
(202) 224-4971
FAX: (202) 224-6163
URL: http://energy.senate.gov

Worldwatch Institute
1776 Massachusetts Ave., NW
Washington, DC 20036-1904
(202) 452-1999
FAX: (202) 296-7365
E-mail: worldwatch@worldwatch.org
URL: http://www.worldwatch.org

RESOURCES

The Energy Information Administration (EIA) of the U.S. Department of Energy (DOE) is the major source of energy statistics in the United States. It publishes weekly, monthly, and yearly statistical collections on most types of energy, available in libraries and online at http://www.eia.doe.gov. The *Annual Energy Review 2000* (2001) provided a complete statistical overview, while the *Annual Energy Outlook 2002* (2001) projected these findings into the future. The EIA's *International Energy Annual 2000* (2002) presents a statistical overview of the world energy situation, while *U.S. Crude Oil, Natural Gas, and Natural Gas Liquids Reserves 2000* (2001) discusses reserves of coal, oil, and gas. The DOE's Office of Policy provides information on the National Energy Policy Plan in *Report of the National Energy Policy Development Group* (2001).

The DOE/EIA also provide *U.S. Nuclear Reactors 2002, Natural Gas Annual 2000* (2001), *Renewable Energy Annual 2000* (2001), *Petroleum: An Energy Profile* (1999), *Electric Power Annual 2000 Volume 1* (2001), and *Natural Gas Annual 2000* (2001). The DOE/EIA also make available information on the Waste Isolation Pilot Project, the development of alternative vehicles and fuels, and electric industry restructuring.

The U.S. General Accounting Office (GAO) has published numerous reports, including *Energy Markets: Results of FERC Outage Study and Other Market Power Studies* (2001), *Nuclear Safety: The Convention on Nuclear Safety* (1999), *Nuclear Cleanup: DOE Should Reevaluate Waste Disposal Options Before Building New Facilities* (2001), *Nuclear Waste: Technical, Schedule, and Cost Uncertainties of the Yucca Mountain Repository Project* (2001), and *Nuclear Waste: Uncertainties About the Yucca Mountain Repository Project* (2002). Also useful is *Nuclear Safety: International Assistance Efforts to Make Soviet-Designed Reactors Safer* (1994) and *Nuclear Safety: Concerns with the Continuing Operation of Soviet-Designed Nuclear Power Reactors* (2000).

The U.S. Department of Transportation's Bureau of Transportation Statistics provides transportation information in its *Transportation Statistics Annual Report 2000* (2001).

The U.S. Environmental Protection Agency (EPA) maintains an Internet home page (http://www.epa.gov/radiation/yucca/index.html) for the Yucca Mountain Repository. Information from that site was extremely useful in the updating of this book. The EPA also provides *Light-Duty Automotive Technology and Fuel Economy Trends 1975—2001* (2001) and *Inventory of U.S. Greenhouse Gas Emissions and Sinks: 1990—1998* (2000).

The U.S. Nuclear Regulatory Commission (NRC) and the United Nations are also important sources of information. The NRC provides the document *NRC—Regulator of Nuclear Safety*. The United Nations Environmental Programme (UNEP) provides *Climate Change Information Kit* (2001), a document based on the work of the Intergovernmental Panel on Climate Change (IPCC).

INDEX

natural gas, 49*f*, 53, 58(*f*3.12), 60*f*
   petroleum, 8(*f*1.7), 30*t*-31*t*, 33-34
India, 100
Industrial energy consumption, 11, 15*t*
   coal, 66, 67(*f*4.7)
   electricity, 124*t*
   natural gas, 49*f*, 53, 54*t*-55*t*, 57*t*
Industrial Revolution, 1
International Energy Outlook, 99, 135
International energy usage, 18-19
   coal, 18*t*, 69, 71*f*
   conservation efforts, 149
   consumption by country, 20*t*-21*t*
   electricity, 125, 128*t*-129*t*
   Generation IV International Forum, 73-74
   geothermal, 18*t*, 97-98
   hydroelectric power, 18*t*, 95
   Kyoto Protocol, 136-138
   Middle East, 32-33
   natural gas, 53, 56, 59(*f*3.14)
   nuclear energy, 79-80
   obstacles toward renewable sources, 102
   petroleum, 40-45
   production by source, 18*t*
   reserves of energy, 112-117
   solar power, 100-101
   top energy producing countries, 19(*f*1.11)
   wind, 98-99
   See also Renewable energy sources
Iranian revolution, 2
Iraq, 3, 7, 37
   See also Middle East
Isotopes, 75

J

Japan, 79, 83, 102

K

Kenya, 100
Kuwait, 3, 7, 37
   See also Middle East
Kyoto Protocol, 136-138

L

Landfill gas recovery, 94
Lawn equipment, 149
Lease condensate, 26, 51*f*
Legislation and international treaties
   Alaska National Interest Lands
      Conservation Act (1980), 108
   Automobile Fuel Efficiency Act (1975),
      139
   Clean Air Act (1990), 67, 69, 142
   Clean Coal Technology Law (1984), 67
   Comprehensive Electricity Competition
      Act (1999), 125
   Energy Policy Act (1992), 98
   National Appliance Energy Conservation
      Act (1987), 147-148
   Natural Gas Policy Act (1978), 53
   Oil Pollution Act (1990), 38
   Price-Anderson Act (1987), 84
   Public Utilities Regulatory Policies Act
      (1978), 89
   Uranium Mill Tailing Radiation Control
      Act (1978), 85

Lignite, 62
   See also Coal
Liquefied natural gas, 143
   See also Natural gas
Liquefied petroleum gas, 142

M

Magma resources, 96
Market share by vehicle type, 139(*f*9.4)
Mass burn system, 94
Methane, 47, 90, 93
Methanol, 142
Metzamor, Armenia, 83
Middle East, 32-33
   Iraq, 3, 7, 37
   Kuwait, 3, 7, 37
   reserves of natural gas, 114
   reserves of oil, 112
   Saudi Arabia, 2, 18
   use of pipelines in, 37
Minami, Japan, 83
Mining methods for coal, 63-64, 65*f*
Mining of uranium, 75-76

N

Nader, Ralph, 81
National Appliance Energy Conservation
   Act (1987), 147-148
National Energy Policy Plan, 150
Natural gas, 1, 47-60
   commercial energy consumption of, 13*t*
   consumption and production flow, 7*f*, 48*f*
   consumption of, 53-56, 54*t*-55*t*, 59*f*
   electric power sector consumption of, 17*t*
   field counts, 51*f*
   future trends for, 56
   imports and exports of, 9*t*-10*t*, 60*f*
   industrial energy consumption of, 15*t*
   international production, 18*t*
   international reserves of, 113-114, 116*t*-
      117(*t*7.8)
   international usage, 53, 56
   plant liquids, 26
   prices, 53
   prices for, 12*t*
   producing energy, 4*t*-5*t*, 6
   production of, 47-51, 59(*f*3.15)
   reserves of, 48, 51, 106-112
   residential energy consumption of, 14*t*
   supply and disposition, 49*f*
   transmission of, 51-53
   transportation energy consumption of,
      16*t*
   underground storage, 50-51, 52(*f*3.9)
   use by sector, 11
Natural Gas Policy Act (1978), 53
Nevada, 86, 87*f*
New Mexico, 86
New Zealand, 98
Nonassociated (NA) natural gas, 56
   See also Natural gas
Norway, 102
Nuclear energy, 73-88
   aging power plants, 80
   chain reaction, 75*f*
   consumption and production flow, 7*f*

domestic usage, 77-79
electric power sector consumption of,
   17*t*, 119
explanation of, 74-75
fuel cycle, 74(*f*5.2)
fusion, 84-85
future of, 83-84
generating electricity, 11, 128*t*-129*t*
international generation of, 79*t*
international production, 18*t*, 79-80
mining uranium, 75-76
operable plants in United States, 77*f*, 78*t*
plant operations, 76*t*
production of, 76-77
radiation sources, 74(*f*5.3)
radioactivity, 75
safety concerns, 73, 80-83
Soviet-built reactors, 82-83, 82*f*
spent nuclear fuel, 88*t*
waste disposal, 85-88
Nuclear Regulatory Commission (NRC), 83-
   84
Nuclear Waste Policy Act (1982), 86

O

Ocean energy, 101-102
Ocean thermal energy conversion (OTEC),
   101, 102
Offshore drilling, 47-48, 50(*f*3.7), 51*f*
   See also Drilling
Oil
   embargo of 1970s, 1-2, 89
   industry, 38
   tankers, 38
   wells, 25-26
   See also Petroleum
Oil & Gas Journal, 112
Oil Pollution Act (1990), 38
Organization of Petroleum Exporting
   Countries (OPEC), 2
   consumption of oil, 43*t*-44*t*
   oil refineries in, 28
   prices for petroleum, 34
   production of oil, 20-21, 41*t*-42*t*
Ozone depletion, 146

P

Passive solar energy systems, 99*f*
Peach Bottom, PA, 81
Pennsylvania, 73, 76, 77, 80-81
Per capita energy use, 150(*f*9.8)
Persian Gulf War, 2*f*, 3, 37, 139
Petroleum, 1, 25-45
   amount of energy produced by, 4*t*-5*t*, 6
   commercial energy consumption of, 13*t*
   conservation issues, 37
   consumption of, 7*f*, 11, 43*t*-44*t*
   countries producing, 40*f*
   crude oil reserves, 105, 106*f*, 107*f*
   drilling for oil, 25-26
   electric power sector consumption of, 17*t*
   embargo of 1970s, 1-2, 89
   environmental issues, 37-38
   events affecting cost of, 2*f*
   future trends in, 45
   imports and exports of, 9-10, 10*t*
   industrial energy consumption of, 15*t*

Thermal reservoirs, 95
Thermochemical conversion, 90
Three Mile Island, PA, 73, 76, 77, 80-81
Tidal power, 102
Tokamak Fusion Test Reactor (TFTR), 84-85
Transportation, 11, 16*t*
   air travel, 146-147
   automobile efficiency, 138-142
   conservation options, 138-147, 138*t*
   fuel efficiency, 150(*f*9.9)
   by mode, 23(*f*1.13)
   natural gas, 49*f*, 54*t*-55*t*, 57*t*
Transportation Statistics Annual Report, 138
Transuranic wastes (TRU), 86

U
Ukraine, 73, 81-82
Underground mining, 63, 65*f*
United Nations' Intergovernmental Panel on Climate Change (IPCC), 136

United States
   coal production in, 64-66
   locations of coal deposits in, 62-63
   oil refineries in, 28
   operable nuclear power plants in, 77*f*, 78*t*
   petroleum consumption, 20-21
   See also Domestic energy usage;
     Domestic production; States
*The U.S. Coal Industry in the 1990s: Low
   Prices and Record Production,* 68
Uranium, 74
   mill tailings, 85
   mining, 75
   reserves of, 112, 114, 116(*t*7.7)
Uranium Mill Tailing Radiation Control Act
   (1978), 85

W
Waste, 93-94, 103
   electric power sector consumption of, 17*t*
   industrial energy consumption of, 15*t*

Waste alcohol, 4-5*t*
Waste disposal of nuclear energy, 85-88, 85*f*
Waste Isolation Pilot Plant (WIPP), NM, 86
Watt, James, 119
Wave energy, 102
Wildlife, 108, 109
Wind energy, 98-99
   amount of energy produced by, 4*t*-5*t*
   electric power sector consumption of, 17*t*
Wood, 1
   amount of energy produced by, 4*t*-5*t*
   commercial energy consumption of, 13*t*
   electric power sector consumption of, 17*t*
   industrial energy consumption of, 15*t*
   residential energy consumption of, 14*t*
World Oil (journal), 112
Worldwide energy usage
   See International energy usage

Y
Yucca Mountain, Nevada, 86, 87*f*